AN OUTLINE FOR DENTAL ANATOMY

An Outline for Dental Anatomy

DOROTHY PERMAR, B.S., M.S.

Professor of Dentistry, College of Dentistry
The Ohio State University, Columbus, Ohio

LEA & FEBIGER · 1974 · PHILADELPHIA

Library of Congress Cataloging in Publication Data

Permar, Dorothy.

    An outline for dental anatomy.

1. Teeth—Outlines, syllabi, etc.   2. Head—Outlines, syllabi, etc.
I. Title. [DNLM: 1. Jaw—Anatomy and histology. 2. Mouth—
Anatomy & histology. 3. Tooth—Anatomy and histology. WU
101 P451o 1974]

RK280.P42 1974   611′.314   73-5884

ISBN 0-8121-0455-2

Library of Congress Catalog Card Number 73-5884

Published in Great Britain by Henry Kimpton Publishers, London

Printed in the United States of America

Preface

An Outline for Dental Anatomy is intended primarily to guide the student in the study of tooth morphology and to help him understand the relationship of teeth to one another and to the bones, muscles, and nerves closely associated with the dentition.

Beginning students have dissimilar backgrounds. This book is planned to teach to the level of those who have had no formal course in head anatomy but who have studied mammalian anatomy and are acquainted with the general structure of the head. Students with no knowledge at all of the head will need to refer to a book devoted to human anatomy.

Included here are descriptions of external and internal tooth morphology, and of the skull, the temporomandibular articulation and muscles of mastication, the nerves and blood vessels associated with the teeth, and occlusion of the teeth.

Every effort needs to be made to help the student study in a way that will be time saving and will make learning a useful and interesting experience. With this object in view this book has been put in outline form, in brief sentences omitting unnecessary words, to give easy reference back and forth from tooth description to tooth specimen. Learning seems to progress more smoothly in this way than by merely reading descriptions of teeth, and there is less tendency for the learner to memorize, for purpose of answering quiz questions, words and sentences that hold little meaning and are soon forgotten. The outline is intended to be used in connection with lecture or discussion periods and laboratory periods.

It is my pleasure to thank Dr. Gary Racey, formerly my teaching associate, now Instructor in Oral Surgery, for writing the chapter on occlusion. To Miss Maureen Shalf I express my gratitude for making most of the drawings in this book. I am indebted to Mr. Ralph Ulbrich, Mr. Norman Marston, and Miss Adele Floridia for assistance with photography; to Mrs. Pamela Yun for careful typing of the manuscript; to Miss Jane Crawford, student in dental hygiene, for a drawing of the mandibular canine tooth; and to Dr. Bernard Snyder for extracted teeth—gallons of them. To the many students who have contributed to the education of fellow students and teachers alike by bringing to classes tooth specimens of special interest, and by reporting research papers they have read and observations they have made, I say thank you for your thoughtfulness.

Dorothy Permar

Columbus, Ohio

Contents

Introduction

Dental anatomy includes the surface form of the oral cavity, the external morphology and internal composition of the individual teeth, the relationship of the teeth to one another and to the skull bones in which they are set, the joints that enable movement of the lower jaw against the upper, the muscles that bring about this movement, and the nerves and blood vessels that supply the muscles and bones and teeth.

For study of surface form of the oral cavity, tooth arrangement, and tooth relationships, you may find some of your best specimens to be the mouths of your family, friends, and classmates. With a little circumspection in asking their cooperation, you will probably find them willing to assist in your education.

For learning tooth morphology, that is, the form of individual teeth, the best possible material is a collection of as many relatively undamaged extracted teeth as you are able to acquire. Tooth models, if available, are useful in learning the distinguishing characteristics of different kinds of teeth; but only by the study of natural teeth can one come to appreciate the variability of the dentition.

Large numbers of nearly intact teeth are sometimes not easy to obtain. Your best source of supply may be a dentist whose clinical practice includes many extractions. If he is presented with a quart jar of formalin, he will remember his own student days and will probably be glad to put extracted teeth in the jar. Do not expect him to sort out the damaged teeth; this is for the student to do.

Tooth specimens should be immersed for days or weeks in 10 per cent buffered neutral formalin,* scraped clean with a knife (soaking for several hours in hydrogen peroxide before scraping is helpful), and stored for preservation in clean 10 per cent neutral formalin.

A good way to ruin your valuable tooth specimens is to use as a preservative formalin which has not been buffered (the subsequently formed formic acid will dissolve the tooth enamel), or to let the specimens dry out (they will become fragile and break).

This book is composed in the form of an outline rather than a narrative to give you easier and quicker reference in your study of the morphology of teeth and the relationship of teeth to their surrounding and supporting tissues. The best method of study with the outline is to sort out a number of teeth of the type you are going to examine—say maxillary central and lateral incisors—and place them and the outline before you on a table. Read the outline, and step by step identify on your several specimens each feature as it is described. Notice the anatomic variations in teeth of the same type. With the tip of your finger, feel the curva-

*Buffered neutral formalin (pH 7)

Formalin, full strength (40 per cent formaldehyde)	100.0 ml
Distilled water	900.0 ml
Sodium phosphate dibasic (anhydrous)	6.5 gm
Sodium phosphate monobasic	4.0 gm

Or, buffer a mixture of 9 parts water and 1 part full strength formalin by adding a few marble chips to prevent the accumulation of formic acid.

tures of the tooth surfaces as you read—sometimes we see as much with our fingers as with our eyes. Working in this way you will find that the descriptions make sense and that you retain the information with little effort. Even if set forth in the most lively language, a detailed description of the morphology of sixteen teeth read as you would read a history book or a mystery story is less conducive to learning than to frustration. But if you become interested in the many consistencies and inconsistencies of tooth anatomy, the study can become a means not only to a sound professional background, but also to an interesting hobby.

One way to reinforce learning of tooth morphology and at the same time develop manual skill is to draw and carve teeth. To draw a tooth accurately (making a poor drawing is of little value) you must examine the tooth until you really *see what you are looking at,* and until you can visualize the relationship of one characteristic to another. To learn to *see what you are looking at* is the objective.

If you carve a tooth from a block of wax, you learn to see the tooth in a third dimension. Also, carving is a manual skill, and this kind of skill is a necessity in almost any phase of dentistry. In the carving exercise you achieve three objectives: you learn the form of the tooth, you learn to see the tooth in 3 dimensions, and you learn to reproduce accurately what you see. You learn to make your fingers create the image that is in your mind.*

External morphology of teeth is not all that is important. Teeth have a hollow center, the pulp cavity, which contains tooth pulp. Tooth pulp is made up of several kinds of connective tissue cells and intercellular substance, throughout which are distributed blood vessels and nerves. Microscopic anatomy of tooth pulp tissue is usually studied in a course in oral histology.

Pulp cavities, their size, shape, and variations are best studied by the interesting operation of grinding off one side of extracted teeth. A dental lathe equipped with a *fine-grained* abrasive wheel about 3 inches in diameter and ½ inch thick can be used to remove any part of the tooth. Simply decide which surface is to be removed, hold the tooth securely in your fingers, and apply this surface firmly to the *flat surface* of the abrasive wheel. Operating the lathe at a fairly high speed is less apt to flip the specimen from your fingers than operating it at a low speed. If you can devise an arrangement by which a small stream of water is run onto the surface of the wheel as the tooth is ground, you will eliminate flying tooth dust and a bad odor of hot tooth tissue. If such an arrangement is not feasible, keep the tooth moist by frequently dipping the surface being ground in water or by dropping water onto the wheel with a medicine dropper. Look often at the tooth surface you are cutting and adjust your applied pressure to attain the plane in which you wish the tooth to be cut.

As you frequently examine the cut surface of the tooth, watch for pits and fissures in the enamel (see the description of occlusal surfaces of posterior teeth) and watch carefully the configuration of any caries lesion as you see it at different levels of grinding. This can be a fascinating pastime as well as a valuable part of your education. You may want to preserve your ground tooth specimens for future examination when, in your course in oral histology, you study the histology of the hard tooth tissues.

For the study of head anatomy it will be much to your advantage to have available a prepared skull. More is learned by holding the skull in your hands and tracing the bone surfaces with your fingers than by energetic attempts to memorize unaccustomed words of uncertain meaning.

Among students who undertake the study of dental anatomy preparatory background differs considerably. If, in your preparatory courses, you have not studied head anatomy, you will need to obtain a book on human anatomy and read about the skull, muscles, nerves, and blood vessels of the head. Study your prepared skull in the same way you study individual teeth: locate the larger bones, and pay particular attention to the maxillary and mandibular bones. Examine the bilateral articulation of the mandible with the cranium (the temporomandibular articulation) and read the outline carefully. The illustrations of the joint should help you to visualize its construction in living persons and the movement of the joint in function.

---

*Directions for drawing and carving teeth are to be found at the end of the book.

If you are so fortunate as to have several skulls available, examine as many as you can and notice the variation in the size and shape of the bones. Pay particular attention to the bones of the upper and lower jaws. You probably will conclude that truly, variability is consistency.

Examine skulls also for the occlusion of the teeth, i.e., the contact of the maxillary and mandibular teeth when the mandible is in different functional positions. It is well to keep in mind, if you see some unexplainable tooth relationships, that sometimes in prepared skulls missing or damaged teeth have been replaced by teeth from other sources. This substitution makes little difference in the study of skull anatomy unless you happen to be studying occlusion.

Judicious note-making during your examination of a number of any kind of specimen will bring into focus features characteristic of the group. For instance, in examining a certain type of tooth, list the characteristics you find constant in all of the teeth in that group, and the characteristics that are constant in many but that vary in a few specimens. And make record of teeth with characteristics so greatly different from all of the other teeth that you regard those teeth as anomalies.

AGAIN YOU ARE URGED TO HAVE TOOTH SPECIMENS BEFORE YOU AS YOU STUDY THE MORPHOLOGY OF TEETH. THIS METHOD IS THE ROAD, PERHAPS THE ONLY ROAD, TO RAPID SUCCESS.

1

The Appearance of the Oral Cavity

UPON LOOKING INTO THE ORAL CAVITY you can see*

1. Lips (entrance)
2. Teeth
3. Tongue
4. Roof of mouth
5. Floor of mouth
6. Cheeks
7. Gingiva
8. Vestibules

LIPS

The lips are redder in young persons than in older ones. In some individuals the lip color is reddish brown due to the presence of brown pigment.

TEETH

There are 2 *dental arches* (rows of teeth): maxillary and mandibular (Figs. 1-1 and 1-2).

Number of teeth:
Number depends on the age of the individual, normal development of the teeth, and number of teeth lost by disease or accident.

The complete primary dentition consists of 20 teeth.

The complete permanent dentition consists of 32 teeth.

At certain ages (from approximately 6 to 12 years) a *mixed dentition* (some primary teeth and some permanent teeth) is present.

TONGUE

The tongue is broad and flat in shape. The shape changes with functional movement.

The *dorsum* (top side) of the tongue is grayish red and is rough. It is covered by several kinds of papillae.

Small papillae are over the entire dorsum.

*You should look into someone's mouth as this description is read and locate each structure mentioned. Use a tongue depressor to retract the lips and cheeks.

Large *circumvallate papillae* form a V-shaped line on the dorsum in the posterior part.

The undersurface of the tongue is shiny and blood vessels are visible.

The *lingual frenum* (frē'num) is a thin sheet of tissue which attaches the center of the undersurface of the tongue to the floor of the mouth.

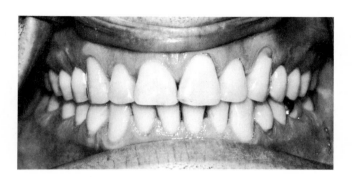

FIG. 1-1 Maxillary and mandibular teeth of the permanent dentition are in occlusion; the labial surfaces of the anterior teeth and the buccal surfaces of the posterior teeth are seen. The gingiva surrounds the teeth and forms the interdental papillae between them. The mandibular vestibule is exposed. Notice how each tooth is in contact with the adjacent teeth; how the maxillary arch overlaps the mandibular arch; and how the greater width of the maxillary central incisors causes each of them to overlap not only the mandibular central incisor but also half of the mandibular lateral incisor. (See Chapter 17 on occlusion.)

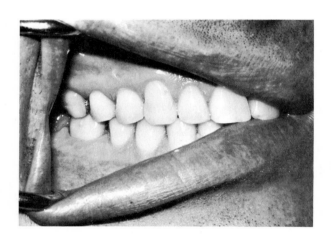

FIG. 1-2. The subject is the same as in figure 1-1, but here the right cheek is drawn back. Notice the arrangement of the maxillary teeth relative to the mandibular teeth: each maxillary tooth overlaps 2 mandibular teeth. There is a clear line between the light-colored gingiva and the darker alveolar mucosa. (See p. 4.)

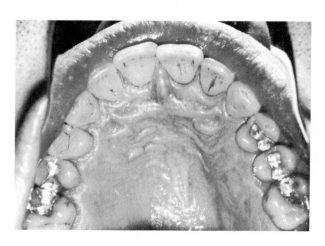

FIG. 1-3. The hard palate is bordered by the maxillary dental arch. The teeth in the picture are, on either side of the arch, the central incisor, lateral incisor, canine, first premolar, second premolar, and first molar. See figure 2-1. The posterior parts of the arch, and the soft palate, are not in the picture. The *palatine rugae* are prominent, as is the *incisive papilla* (the enlargement between the right and left central incisors). The *palatine raphe* is seen but is not conspicuous in this subject. In some mouths the raphe, extending anteroposteriorly in the center of the hard palate, is more easily seen.

ROOF OF THE MOUTH

Hard palate (the anterior part of the roof of the mouth). ~~front~~

Soft palate (the posterior part of the roof of the mouth). back

THE HARD PALATE (Fig. 1-3)

The color is grayish red.

Palatine raphe (pronounced *rā′fē*) is the slightly elevated center line running antero-posteriorly in the hard palate. This is the line of union of the right and left maxillae. Here the bone is immediately under the mucosa covering the palate.

The sides of the hard palate are less hard then the palatine raphe because here there is fat or salivary gland tissue beneath the surface tissue.

Palatine rugae
Rugae (plural, pronounced *roo′jē;* singular is *ruga, rōō′ga*) are more distinct in young persons than in older persons. They are a series of elevations, or wrinkles, running from side to side behind the maxillary anterior teeth.

The end of the hard palate is opposite the third molars.

THE SOFT PALATE

The soft palate is sometimes redder than the hard palate. Its anterior border extends between the right and left third molars.

There is no bone beneath the surface. (Say *AHHHH* and look.)

Uvula (*u′vu la.* Both *u*'s are pronounced like the *u* in *use*) is a small fleshy structure hanging from the center of the posterior border of the soft palate.

Behind the soft palate is the pharynx.

Foveae palatinae (*fŏ′vè ē păl′à tĭn′ē;* L. *fovea,* pit) are a pair of pits in the soft tissue located on either side of the center line near the posterior border of the hard palate. They are openings of ducts of a number of palatine glands.

FLOOR OF THE MOUTH

Beneath the tongue are bilateral bulges caused by the presence there of large salivary glands.

Sublingual folds, called the *plica sublingualis* (*plĭ′kà sub lĭng guál ĭs*), extend horizontally to the right and left of the lower end of the *lingual frenum* where the broad underside of the tongue is attached to the floor of the mouth. Along these folds are openings of ducts from underlying salivary glands located in this region.

At the center line between the right and left sublingual folds is a papilla with a pair of openings from ducts of salivary glands, called *Wharton's duct openings.*

The floor of the mouth is shiny, and some large blood vessels may be seen near the surface.

CHEEKS

The lining of the inside of the cheeks is shiny, and in spots it is rough.

Stensen's duct (the duct from the parotid salivary gland) opens on the inside of either cheek opposite the maxillary first molar. A small elevation can be seen and felt here.

Fordyce's spots are sometimes conspicuous in a horizontal line extending along the inside of the cheek from the corners of the mouth. They are small elevations, sometimes yellow in color, produced by the presence of sebaceous glands—glands of a type otherwise found only in the skin on the outside of the body. Their presence here is often said to be the result of fusion of the upper and lower parts of the cheek during embryonic development. However, such glands have also been found on other parts of the oral mucosa.

GINGIVA

Definition: The soft tissue which surrounds the cervical part of the teeth and is firmly attached to the teeth and to the bone in which the teeth are set (Figs. 1-1 and 1-2).

The gingiva is grayish and stippled in persons with light pigmentation. In persons with dark coloring of the hair and skin, the gingiva may be brown or spotted with brown.

There is a scalloped line between the gingiva and the redder *alveolar mucosa* just below (lower jaw) or above (upper jaw) it.

The *gingival margin* is the occlusal (incisal) border of the gingiva.

The gingival tissue between the teeth is called the *interdental papilla.*

VESTIBULE

The space bounded by the cheek or lip on one side and by the teeth and the gingiva covering the bone of the maxilla or mandible on the other side is called a *vestibule* (maxillary or mandibular) (Fig. 1-2).

The *labial frenum* is the thin sheet of tissue that attaches the center of the lip (upper and lower) to the mucosa covering the jaw.

The *buccal frenum,* in the area of the premolars (maxillary and mandibular), loosely attaches the cheek to the mucosa of the jaw. These frenums may be seen by pulling the lower lip out and down, and the upper lip out and up.

ORAL MUCOUS MEMBRANE (ORAL MUCOSA)

A *mucous membrane* lines any body cavity that opens to the outside of the body. The *oral mucous membrane* lines the oral cavity. It resembles the skin on the outside of the body, except that it is more delicate in structure than the skin.

The mucosa of the mouth is a sturdy type of mucous membrane, and in the areas in which it gets most wear it is most sturdy, e.g., the roof of the mouth and the gingiva. Its appearance in these areas of greater wear is grayish rather than red, as it is in the floor of the mouth and cheeks where it is more protected.

Notes

2

Terminology

When we enter upon any new field of study it is necessary at once to learn the particular language of that field. Without adequate vocabulary we can neither understand nor make ourselves understood. Definitions and explanations of terms frequently used in discussing tooth morphology will be helpful in understanding the descriptions of teeth that follow.

THE JAWS

a. MAXILLA—the upper jaw. There are 2 bones to the upper jaw, i.e., 2 maxillae, although the word maxilla is often used in the singular to mean the combined right and left maxillary bones (Fig. 13-1).

b. MANDIBLE—the lower jaw. This is a single bone (Fig. 13-2).

THE DENTITION

The word *dentition* means all of the teeth considered collectively. In human beings there are 2 dentitions:

a. THE PRIMARY DENTITION

This is also called the *deciduous dentition*. It is composed of 20 teeth: 10 maxillary, 10 mandibular. These teeth are lost at the end of a certain period and are replaced by the permanent dentition.

b. THE PERMANENT DENTITION

This is also called the *succedaneous dentition*. It is composed of 32 teeth: 16 maxillary, 16 mandibular.

NAMES OF TEETH AND DENTAL FORMULAS

a. PRIMARY TEETH (Fig. 11-1)

The complete primary dentition has in *each quadrant* of the mouth 2 incisors, 1 canine (cuspid), 2 molars. A *dental formula* briefly indicates this information for the maxillary and mandibular teeth on one side of the mouth.

$$\text{I}\frac{2}{2}\,\text{C}\frac{1}{1}\,\text{M}\frac{2}{2} = 10 \text{ teeth on either side; 20 teeth in all.}$$

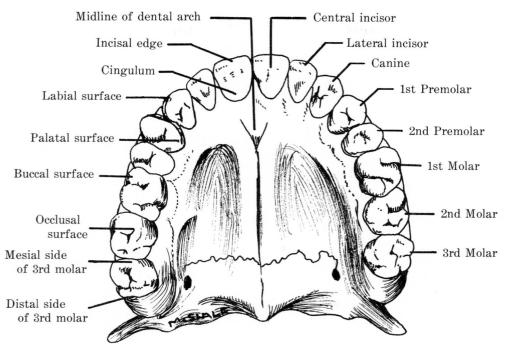

FIG. 2-1. Diagram of the maxillary dental arch and the bones of the hard palate.

b. PERMANENT TEETH (Figs. 2-1 and 2-2)

The complete permanent dentition has in *each quadrant* of the mouth 2 incisors, 1 canine (cuspid), 2 premolars (bicuspids), 3 molars. The dental formula is

$$I\frac{2}{2} \, C\frac{1}{1} \, PM\frac{2}{2} \, M\frac{3}{3} = 16 \text{ teeth on either side; 32 teeth in all.}^{*}$$

WORDS USED TO DESCRIBE TEETH

a. PARTS OF A TOOTH (Figs. 2-3 and 2-4)

Crown

    Anatomic crown—the part of a tooth that has an enamel surface.

    Clinical crown—the part of a tooth that is exposed to the oral cavity.

Root

    Anatomic root—the part of a tooth that has a cementum surface.

    Clinical root—the part of a tooth that is embedded in the jaw and is not exposed to the oral cavity. (See Chapter 12.)

Cervix—the part of the root near the cementoenamel junction (Fig. 2-4).

*For the dental formulas of some animals, see page 54.

IDENTIFICATION OF TEETH BY NUMBER AND TIME OF EMERGENCE INTO ORAL CAVITY*

PERMANENT DENTITION

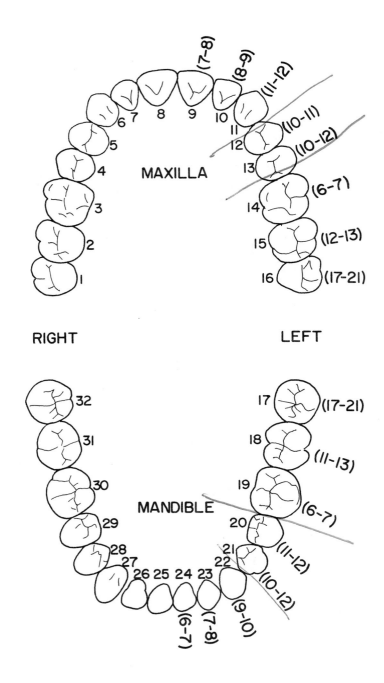

*EMERGENCE TIME IN YEARS INDICATED BY FIGURES IN PARENTHESES.

FIG. 2-2. A diagrammatic illustration of the maxillary and mandibular permanent dental arches. The numbers in parentheses indicate the ages at which each of the teeth may be expected to emerge into the oral cavity. The numbers 1 to 32 inside the arches indicate one of several ways in which teeth are coded (rather than named) for purposes of record-keeping. Starting at the maxillary right (of the subject) third molar, which is number 1, the numbers extend through consecutive teeth to the maxillary left third molar, number 16, then continue from the mandibular left third molar, number 17, to the mandibular right third molar, number 32. See figure 2-1.

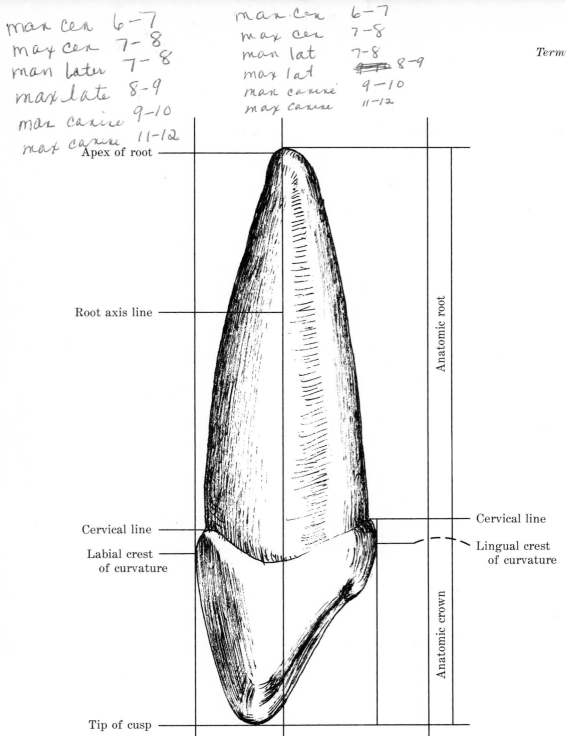

Apex of root

Root axis line

Anatomic root

Cervical line

Cervical line

Labial crest of curvature

Lingual crest of curvature

Anatomic crown

Tip of cusp

FIG. 2-3. Mesial side of a maxillary right canine. Drawing of a tooth model. The root axis line bisects the root in the cervical area. Customarily, other parts of the tooth are located or described relative to this line. In this case, for instance, the cusp tip is labial to the root axis line.

Root Trunk—the part of the root of a *molar* tooth near the cementoenamel junction. It is the part that is not furcated (Figs. 8-1a, 8-1c).

Apex of Root—the tip end of the root (Fig. 2-3).

Enamel—the hard, shiny surface of the anatomic crown.

Cementum—the dull surface of the anatomic root.

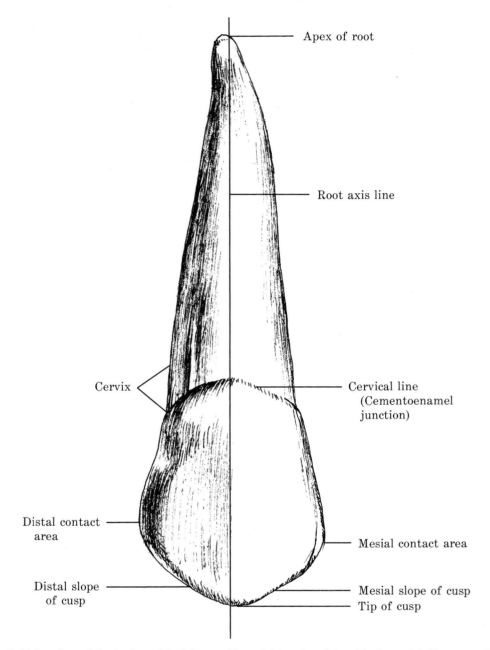

FIG. 2-4.   Labial surface of the tooth model of the maxillary right canine pictured in figure 2-3. The root axis is again determined by bisecting the root at the cervix.

> *Dentin*—the hard tissue underlying enamel and cementum and making up the bulk of the tooth (Fig. 10-1).

> *Cervical Line* (cementoenamel junction)—the line at the cervical border of the anatomic crown where the enamel and the cementum meet (Fig. 2-3).

> *Cusp*—a point, or peak, on the occlusal surface of molar teeth and on the incisal edge of canine teeth (Figs. 2-3, 2-4, and 9-2).

Cingulum (the *c* and *g* are pronounced as in *singular*)—the enlargement on the cervical third of the lingual surface of the crown in anterior teeth (incisors and canines).

Mamelon—one of 3 tubercles sometimes present on the incisal edge of an incisor tooth (Fig. 4-1a).

Cusp Slopes—the inclined surfaces that form an angle at the cusp tip (Fig. 2-3).

Marginal Ridge

On incisor and canine teeth, the ridges on the mesial and distal borders of the lingual surface (Fig. 3-2).

On posterior teeth, the ridges on the mesial and distal borders of the occlusal surface (Fig. 9-2).

Triangular Ridge—on the occlusal surface of posterior teeth, the ridge from any cusp tip to the center of the occlusal surface (Fig. 8-3).

Transverse Ridge—a ridge crossing the occlusal surface in a buccolingual direction made up of the triangular ridges of a buccal and a lingual cusp (Fig. 9-2).

Oblique Ridge—a ridge found on maxillary molar teeth, which crosses the occlusal surface obliquely, made up of the triangular ridges of the mesiolingual and the distobuccal cusps (Fig. 8-3).

Labial Ridge—a ridge running cervicoincisally in approximately the center of the labial surface of a canine tooth.

Buccal Ridge—a ridge running cervico-occlusally in approximately the center of the buccal surface of premolar teeth.

Cervical Ridge—a ridge running mesiodistally on the cervical ⅓ of the crown.

Fossa (plural, *fossae*)—a depression, or hollow, found on the lingual surfaces of some anterior teeth (particularly maxillary incisors), and on the occlusal surfaces of all posterior teeth (Figs. 3-3 and 8-3).

Groove—a sharply defined, narrow, linear depression, short or long, formed during tooth development (not due to damage after tooth eruption). The major grooves are named according to their location (Fig. 6-4).

Fissure—a channel, or ditch, narrow and sometimes deep, formed at the bottom of a groove during the development of a tooth. This can scarcely be considered an abnormality since it occurs so frequently (Fig. 9-2).

Pit—a small, often deep, depression formed at the union of 2 or more grooves; or sometimes at the end of a single groove (as on a mandibular molar at the cervical end of a buccal groove) (Fig. 9-3a).

Supplemental Grooves—small, irregularly placed grooves found usually on occlusal surfaces.

Sulcus—a broad depression on the occlusal surfaces of posterior teeth, the inclines of which meet in a developmental groove (Fig. 6-3).

b. SURFACES OF A TOOTH (Fig. 2-1)

Facial Surface—the surface next to the face; the outer surface when the tooth is in place in the mouth.

Buccal Surface (pronounced like *buckle*)—meaning *cheek*. Another name for the facial surface of the posterior teeth (next to the cheek).

Labial Surface—meaning *lip*. Another name for the facial surface of anterior teeth (next to the lip).

Mesial Surface—the surface of the tooth nearest the midline of the dental arch.

Distal Surface—the surface of the tooth farthest from the midline of the dental arch.

Lingual Surface—the surface of mandibular teeth nearest the tongue.

Palatal Surface—the surface of maxillary teeth nearest the palate. This surface is frequently also called the lingual surface.

Occlusal Surface—the chewing surface of posterior teeth.

Incisal Edge—the cutting edge, or surface, of anterior teeth.

c. POINTS OF REFERENCE

Medial—toward the center line of the *body*.

Mesial—toward the center line of the *dental arch;* i.e., toward the point between the right and left central incisor teeth.
Proximal Surface (*approximal surface*)—the surface of a tooth which is next to an adjacent tooth (either the mesial or the distal surface).

Crest of Curvature—the highest point of a curve. A term usually used with reference to the points on the facial and lingual surfaces of the crown touched by a tangent line drawn parallel to the root axis (Fig. 2-3).

Contact Area (*contact point*)—the small area on the mesial and distal surfaces of a tooth which, when the teeth are in good alignment in the dental arch, touches (is in contact with) the adjacent tooth in the same arch.

Divisions of the Crown of a Tooth (used for purposes of description) (Fig. 2-5):

Divisions Cervico-occlusally (*cervicoincisally*)—are demonstrated by lines drawn *horizontally* on the tooth crown:

      Cervical third
      Middle third
      Occlusal (incisal) third

FOR PURPOSES OF DESCRIPTION TEETH ARE DIVIDED INTO AREAS.

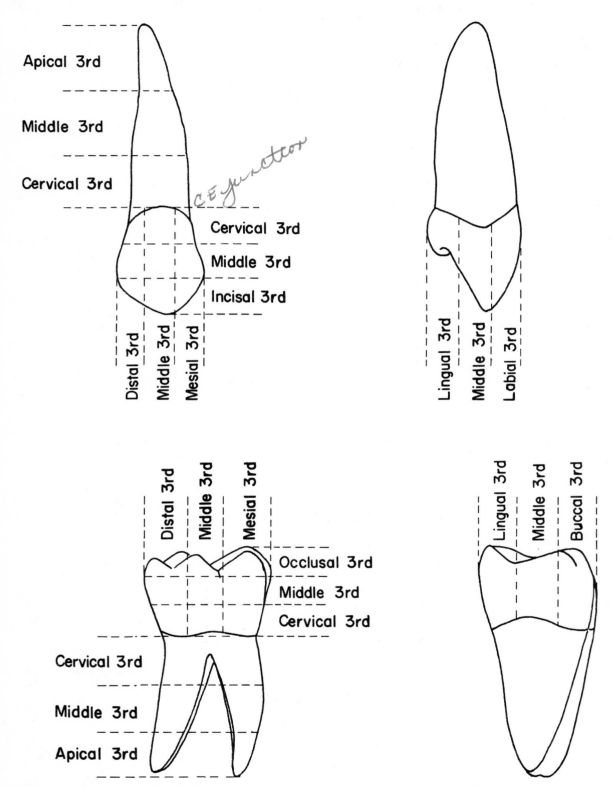

Apical 3rd

Middle 3rd

Cervical 3rd

C E junction

Cervical 3rd

Middle 3rd

Incisal 3rd

Distal 3rd   Middle 3rd   Mesial 3rd

Lingual 3rd   Middle 3rd   Labial 3rd

Distal 3rd   Middle 3rd   Mesial 3rd

Occlusal 3rd

Middle 3rd

Cervical 3rd

Cervical 3rd

Middle 3rd

Apical 3rd

Lingual 3rd   Middle 3rd   Buccal 3rd

FIG. 2-5. Diagram of a maxillary canine and a mandibular first molar to show the manner in which the parts of a tooth may be divided for purposes of description.

Divisions Mesiodistally—are demonstrated by lines drawn *vertically* on the facial or lingual surface of the crown:

> Mesial third
> Middle third
> Distal third

Divisions Faciolingually—are demonstrated by lines drawn *vertically* on the mesial or distal surface of the crown:

> Facial third (labial or buccal third)
> Middle third
> Lingual third

d. COMBINED TERMS (notice spelling)

Mesiodistal	Distoincisal
Mesio-occlusal	Distobuccal
Mesioincisal	Labiolingual
Mesiolingual	Buccolingual
Mesiobuccal	Faciolingual
Distolingual	Cervicoincisal
Disto-occlusal	Cervico-occlusal

EVOLUTION OF TEETH

In terms of the evolution of the dentition, teeth are said to have developed from *lobes*. For example, the mandibular first molar supposedly developed from 5 lobes: it has 5 cusps—3 buccal, 2 lingual. Incisor teeth supposedly developed from 4 lobes: they have 3 mamelons and 1 cingulum (1 lingual lobe). All teeth show evidence of having developed from 4 or more lobes.

Notes

3

Permanent Maxillary Incisors

General Description of Maxillary Incisors

There are 4 maxillary incisor teeth: 2 *central incisors* and 2 *lateral incisors*.

Central Incisors (First Mandibular Incisors) (Fig. 2-1)

The central incisors are located on either side of the maxillary arch with their mesial surfaces in contact. (In some individuals these teeth, and other teeth, are separated by more or less of a space. This space is called a diastema [dī′ à stē′ mà].)

Lateral Incisors (Second Mandibular Incisors) (Fig. 2-1)

The lateral incisors are distal to the central incisors, their mesial surface in contact with the distal surface of the adjacent central incisor.

DEVELOPMENTAL DATA

Teeth develop within the jaws (in the maxillae and mandible). When the crown is completely formed the tooth starts to move occlusally: the maxillary teeth move down; the mandibular teeth move up.

In the following table* the figures given opposite *Beginning of enamel matrix and dentin formation* indicate the age of the individual at the time the designated tooth tissues begin to form.

	Central incisors	*Lateral incisors*
Beginning of enamel matrix and dentin formation	3-4 mos.	10-12 mos.
Enamel completed	4-5 yrs.	4-5 yrs.
Emergence into oral cavity	7-8 yrs.	8-9 yrs.
Root completed	10 yrs.	11 yrs.

The figures opposite *Enamel completed* indicate the age of the individual at the time the formation of the enamel of the crown of the tooth is completed.

*Dates given in Orban, B. J. (Ed.), *Oral Histology and Embryology*, 5th ed., edited by H. Sicher, St. Louis, Mosby, 1962.

Root formation begins after the crown is formed, and at this time the tooth starts its occlusal movement. This tooth movement is called *eruption*. In the process of eruption the tooth crown emerges into the oral cavity. The eruptive movement continues after the incident of emergence and eventually the tooth comes into occlusion with teeth in the opposite arch.

In the table the figures opposite *Emergence into the oral cavity* indicate the age of the individual at the time the tip of the tooth crown emerges through the oral mucosa into the mouth. There is actually some range in emergence time of any type of tooth.

The root ordinarily is not completely formed until from 1 to 4 years after the crown has become visible in the mouth. In the table the figures opposite *Root completed* indicate the age of the individual at the time root formation is completed.

FUNCTIONS

The maxillary incisor teeth function in cutting food and in enabling articulate speech. (Consider the enunciation of a toothless person.) They also help in maintaining a good appearance, for by our standards a person lacking one or more maxillary incisor teeth has an undesirable appearance.

MORPHOLOGY

The morphology, or anatomy, of a tooth is described just as the morphology of a bone is described. A tooth has 5 surfaces, or 4 surfaces and a cutting edge, depending on whether it is a posterior tooth (in the back part of the mouth) or an anterior tooth (in the front of the mouth). A tooth also has a number of characteristic grooves and ridges. In the study of tooth morphology the grooves and ridges on the various surfaces are described and named.

The external morphology of a tooth is customarily described from 5 aspects: (1) facial (i.e., labial or buccal), (2) lingual, (3) mesial, (4) distal, (5) incisal (or occlusal).

Maxillary Central Incisor

In the study of any single type of human tooth, as for instance the maxillary central incisor, it is necessary to understand that this tooth varies in form in different individuals about as much as different individuals vary in general form from one another. A study of a collection of 100 maxillary central incisors would show a considerable difference in such characteristics as size, shape, relative proportions, color.[7] Therefore, *such information as crown length, crown width, and root length are to be regarded only as approximate figures—i.e., usual, or commonly found measurements.*[6]

SIZE (Measured on the labial surface) (Fig. 18-1)

Crown length (cervicoincisal measurement)	10 mm.
Crown width (greatest mesiodistal measurement)	9 mm.
Root length (cervical line to apex)	12 mm.

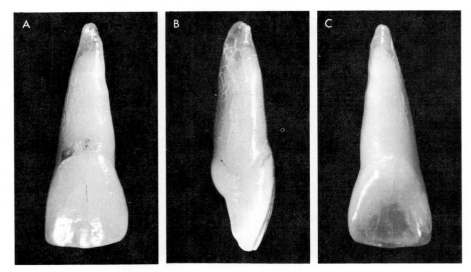

FIG. 3-1.   Maxillary left central incisor. **a.** Labial surface. **b.** Mesial surface. **c.** Lingual surface. In this tooth are a distinct but shallow lingual fossa, and clear but not prominent marginal ridges.

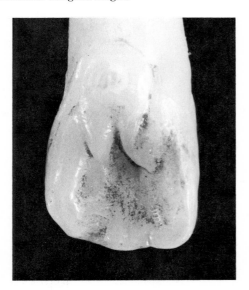

FIG. 3-2.   Lingual surface of a maxillary left central incisor. Compare the lingual fossa, the marginal ridges, and the cingulum with those of the tooth in figure 3-1c.

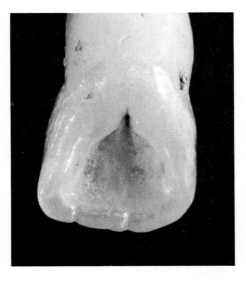

FIG. 3-3.   Lingual surface of a right maxillary central incisor. This tooth, and the tooth in figure 3-2, are clearly of the type called *shovel-shaped*. Notice the pit at the incisal edge of the cingulum—the sort of place caries can develop and penetrate without being quickly noticed.

LABIAL ASPECT (Examine several extracted teeth as you read. Hold maxillary teeth root up and crown down, the position of the teeth in the mouth.)

1. CROWN (Fig. 3-1a)

    For purposes of description, a tooth crown is customarily divided into thirds vertically and horizontally (Fig. 2-5).

    Shape

    The crown of the maxillary central incisor is relatively narrow in the cervical third; it becomes broader toward the incisal third. (Move your finger over the sides of the crown from the cervical line to the incisal edge.)

    The mesioincisal angle is slightly rounded; the distoincisal angle is more rounded.

    The distal side of the crown is more convex than the mesial.

    Contact areas

    Mesial: In the incisal third, near the mesioincisal angle.

    Distal: Near the junction of the incisal and middle thirds.

    Examine the teeth of several persons and notice the location of the contact areas. Sometimes the distal contact area is in the middle third.*

    Morphology of the labial surface

    Mamelons are usually seen on newly emerged incisor teeth. They are the 3 scallops on the incisal edge. Usually they are worn off after the tooth comes into functional position. (If you have a 6-year-old friend, make him smile.)

    Developmental depressions are the 2 very shallow vertical depressions which divide the labial surface into 3 areas. These 3 areas are sometimes called the mesial, middle, and distal *lobes*.

    Perikymata (pronounced *pĕr·ĭ·kĭ′·mat·ah*) are the very fine horizontal lines on the crown surface. Examine teeth and notice that these lines are closer together in the cervical part of the crown than they are nearer the incisal edge.

    Perikymata are found on the enamel of all teeth. They are most easily seen on the labial surfaces of the anterior teeth because of their accessible location. Perikymata are more prominent on the teeth of young persons than on the teeth of older persons.

---

* Note: It should be understood that the location of contact areas for any tooth may vary slightly among different individuals. Generally speaking, contact areas are located in 1 of 3 places: in the incisal (occlusal) third, at the junction of the incisal and middle thirds, or in the middle third of the crown. The locations of contact areas indicated in this study may be regarded as usual locations, but not necessarily as "typical" or the "most normal." Such terms as *typical* and *normal* have little meaning in the study of tooth morphology.

2. ROOT

For purposes of description, it is customary to divide the root of a tooth into thirds: cervical third, middle third, and apical third (Fig. 2-5).

The root of the maxillary central incisor is thick in the cervical third and narrows through the middle to a blunt apex.

The root is only slightly convex on the labial surface.

LINGUAL ASPECT

1. CROWN (Figs. 3-1c, 3-2, 3-3)

The mesial and distal sides of the crown converge lingually, making the lingual surface narrower than the labial surface.

The *cingulum* on this tooth is usually well developed. It is not in the center of the lingual surface, but lies distal to the line that bisects the root longitudinally in the cervical third (the *root axis line*).*

The *mesial* and *distal marginal ridges* are of varying prominence on different teeth.

The mesial marginal ridge is longer than the distal due to the somewhat distal position of the cingulum.

The *lingual fossa* is developed to varying degrees: in some teeth it is very shallow, in others very deep.

Maxillary incisors with a deep lingual fossa and prominent mesial and distal marginal ridges are sometimes called "shovel-shaped incisors."[1,2,4,5,9]

Other variations of the lingual surface are:

a. The fossa may be deep but smooth, i.e., no lingual ridges.

b. *Lingual ridges,* if present, extend from the cingulum toward the center of the fossa. These ridges may be 1, 2, or 3 in number.

c. There may be no lingual ridges, but a *lingual pit* at the incisal border of the cingulum where the mesial and distal marginal ridges come together in this location (Fig. 3-3).

2. ROOT

The lingual surface of the root is narrower than the labial surface.

The root is flattened on the mesial side of the lingual surface.

---

*The *root axis line* on the facial and lingual surfaces divides the root *at the cervix* into mesial and distal halves. The root axis line on the mesial and distal surfaces divides the root *at the cervix* into facial and lingual halves. See figures 2-3 and 2-4.

MESIAL ASPECT

1. CROWN (Fig. 3-1b)

The crown from the mesial side resembles a wedge.

The incisal edge and the root apex are often nearly centered on the root axis line. Sometimes the incisal edge is slightly labial to the root axis line.

The crest of curvature of the labial surface is usually just incisal to the cervical line.

The crest of curvature of the lingual surface is on the cingulum.

The labial surface of the crown is deeply convex in the cervical third; it is broadly curved, sometimes nearly flat, in the middle and incisal thirds.

The lingual surface is convex over the cingulum and concave from the cingulum nearly to the incisal edge.

The cervical line curves incisally on the mesial surface of the tooth. The curvature is greater on the mesial surface than on the distal surface, extending incisally about $\frac{1}{3}$ of the crown length.

2. ROOT

The root is tapered to the apex.

The lingual surface is nearly straight in the cervical third, then curves labially in the middle and apical thirds.

The labial surface is very slightly convex.

DISTAL ASPECT

1. CROWN

The crown from the distal side resembles a wedge.

The incisal edge is on the root axis line or slightly labial to it; but the distoincisal angle may be lingual to the axis line due to a *very slight* distolingual twist sometimes found in the incisal part of the crown.

The curvature of the cervical line is less on the distal surface than on the mesial.

2. ROOT

The shape of the root is similar to that from the mesial aspect.

INCISAL ASPECT

To follow this description a maxillary incisor tooth should be held in such a position that the incisal edge is toward the observer, the labial surface is at the top, and the observer is looking *exactly along the root axis line.*

The labial surface of the crown usually appears broadly convex. In some teeth it is nearly flat. Occasionally the labial surface of this tooth will appear slightly concave from this aspect, but this is an unusual shape.

The mesial marginal ridge is longer than the distal marginal ridge. This difference in the length of the marginal ridges is, of course, associated with the position of the cingulum distal to the center of the crown.

The incisal edge is slightly curved, the convexity being on the labial side.

The position of the distoincisal angle slightly lingual to the position of the mesioincisal angle sometimes gives the incisal edge a slight distolingual twist.

Maxillary Lateral Incisor

SIZE (Measured on the labial surface)

Crown length (cervicoincisal measurement)	9.3 mm.
Crown width (greatest mesiodistal measurement)	6.4 mm.
Root length (cervical line to apex)	12.7 mm.

LABIAL ASPECT

1. CROWN (Fig. 3-4)

   Shape

   The crown of the lateral incisor is narrower than the crown of the central incisor.

   The mesioincisal and distoincisal angles are usually more rounded on the lateral incisor than on the central incisor.

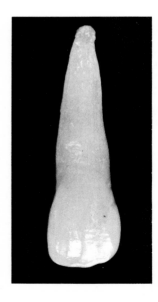

FIG. 3-4. Maxillary left lateral incisor. Labial surface.

Contact areas

> Mesial: In the incisal third.
>
> > (Variation: Sometimes at junction of incisal and middle thirds.)
>
> Distal: At junction of incisal and middle thirds.
>
> > (Variation: Sometimes in middle third.)
>
> Examine the teeth of several persons and notice the location of the contact areas.

Morphology of the labial surface

> The labial surface is very much like that of the central incisor, but usually it is more convex mesiodistally.
>
> Usually the mamelons and labial depressions are less prominent than on the central incisor.

2. ROOT

> The root tapers to the apex, and the apical end is often slightly bent distally.

LINGUAL ASPECT

1. CROWN

> The lingual surface is narrower than the labial surface.
>
> The cingulum is narrower than on the central incisor, and it is almost centered on the root axis line.
>
> The mesial and distal marginal ridges are developed to varying degrees: sometimes they are prominent, sometimes inconspicuous.
>
> The mesial marginal ridge is nearly straight; the distal is slightly convex.
>
> The lingual fossa in some teeth is very shallow, in others deep.
>
> There may be lingual ridges or a lingual pit in the fossa.

2. ROOT

> The root is narrower on the lingual side than on the labial.

MESIAL ASPECT

1. CROWN

> The crown is wedge-shaped.
>
> The curvature of the cervical line is deep. It is deeper on the mesial surface than on the distal.
>
> The incisal edge is on the root axis line, or slightly labial to it.

On the labial surface the crest of curvature is near the cervical line.

On the lingual surface the crest of curvature is on the cingulum.

2. ROOT

The root is proportionally longer than that of the central incisor.

It often has a shallow longitudinal depression on the mesial surface.

It tapers apically.

DISTAL ASPECT

1. CROWN

The crown resembles a wedge.

The cervical line is less curved on the distal side than on the mesial.

2. ROOT

The distal surface of the root is similar to the mesial surface, except that usually there is no depression.

INCISAL ASPECT

The lateral incisor in general resembles the central incisor from this aspect. In the lateral incisor the cingulum is nearly on the center line of the crown.

The labiolingual measurement of the tooth crown is ordinarily less than the mesiodistal measurement, but it is relatively greater in proportion than in the central incisor.

The labial surface of the crown is more convex than that of the central incisor.

Variations in Maxillary Incisor Teeth

Racial differences in the maxillary incisor teeth have been reported in the literature. For example, a high incidence of *shovel-shaped incisors* has been observed in Mongoloid peoples, including many groups of American Indians.[1,3,4] Caucasian and Negro peoples are reported to have less frequent occurrences of this characteristic. Shovel-shape is the term commonly used to designate incisor teeth which on their lingual surfaces have prominent marginal ridges and a deep fossa (Figs. 3-2 and 3-3).

A study of the skulls of American Indians who lived in Arizona about 1100 A.D. has disclosed the occurrence of incisor teeth which have a mesial marginal ridge on the *labial* surface and a depression, or concavity, on the mesial part of the labial surface just distal to this ridge.[8] In these teeth the distal part of the labial surface is rounded in an unusual manner. Such teeth have been referred to as "three-quarter double shovel-shaped"—a descriptive, if ponderous, term.

Labial "shoveling" has been reported also in some Eskimo peoples.

References

1. Carbonelli, V. M., *Variations in the Frequency of Shovel-Shaped Incisors in Different Populations*. In Brothwell, D. R. (Ed.), *Dental Anthropology*, London, Pergamon Press, 1963, pp. 211-234.
2. Brabant, H., *Comparison of the Characteristics and Anomalies of the Deciduous and the Permanent Dentitions*. J. Dent. Res., *46*:897-902, 1967.
3. Dahlberg, A. A., *The Dentition of the American Indian*. In Laughlin, W. S. (Ed.), *The Physical Anthropology of the American Indian*, New York, The Viking Fund, Inc., 1949.
4. De Voto, F. C. H., *Shovel-Shaped Incisors in Pre-Columbian Tastilian Indians*. J. Dent. Res. *50*:168, 1971.
5. De Voto, F. C. H., Arias, N. H., Ringuelet, S., and Palma, N. H., *Shovel-Shaped Incisors in a Northwestern Argentine Population*. J. Dent. Res. *47*:820, 1968.
6. Goose, D. H., *Variability of Form of Maxillary Permanent Incisors*. J. Dent. Res., *35*:902, 1956.
7. Hanihara, K., *Racial Characteristics in the Dentition*. J. Dent. Res., *46*:923-926, 1967.
8. Snyder, R. G., *Mesial Marginal Ridging of Incisor Labial Surfaces*. J. Dent. Res., *39*:361, 1960.
9. Taylor, R. M. S., *Variations in Form of Human Teeth: I. An Anthropologic and Forensic Study of Maxillary Incisors*. J. Dent. Res., *48*:5-16, 1969.

Notes

It is interesting to keep a record of some of your observations as you examine an assortment of teeth. Among items of significance might be the following:

Do you find any maxillary incisors with a pit at the incisal border of the cingulum?

If so, is there any evidence of caries in the pit?

How many teeth have a deep lingual fossa and prominent marginal ridges?

How many maxillary incisors have the incisal edge labial to the root axis line (as seen from the mesial or distal aspect)?

How many have the incisal edge *on* the root axis line?

A sketch of any anomalies is of interest.

4

Permanent Mandibular Incisors

General Description of Mandibular Incisors

There are 4 mandibular incisor teeth: 2 *central incisors*, and 2 *lateral incisors*.

Central Incisors (First Mandibular Incisors)

The mandibular central incisors are on either side of the mandibular arch with the mesial surfaces of the right and left teeth in contact at the midline.

Lateral Incisors (Second Mandibular Incisors)

The mandibular lateral incisors are just distal to the central incisors, with their mesial surface in contact with the distal surface of the adjacent central incisor.

DEVELOPMENTAL DATA

	Central incisors	*Lateral incisors*
Beginning of enamel matrix and dentin formation	3-4 mos.	3-4 mos.
Enamel completed	4-5 yrs.	4-5 yrs.
Emergence into oral cavity	6-7 yrs.	7-8 yrs.
Root completed	9 yrs.	10 yrs.

FUNCTIONS

The mandibular incisors function with the maxillary incisors in cutting food, in production of distinct speech, and in maintenance of a good appearance.

Mandibular Central Incisor

SIZE (Measured on the labial surface)

Crown length (cervicoincisal measurement)	8.5 mm.
Crown width (greatest mesiodistal measurement)	5.5 mm.
Root length (cervical line to apex)	12.0 mm.

The mandibular central incisors are the smallest teeth in the mouth.

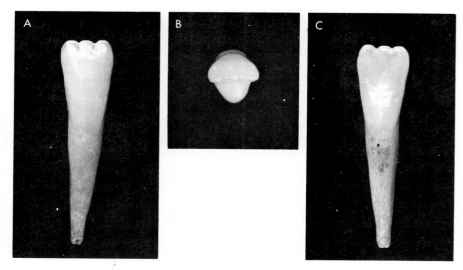

Fig. 4-1a. Mandibular right central incisor. Labial surface. The mamelons are unworn. The tip of the root may be resorbed or, more likely, its formation was not quite completed when its owner lost it. Notice the opening of the root canal at the root tip. **b.** Incisal aspect. The labial side is toward the top. **c.** Lingual surface. The crown surface is concave and nearly smooth.

LABIAL ASPECT (Examine several extracted teeth as you read. Hold mandibular teeth root down and crown up, the position of the teeth in the mouth.)

1. CROWN (Fig. 4-1a)

    Shape

    The crown is about $\frac{2}{3}$ as wide as the crown of the maxillary central incisor.

    > Examine someone's mouth and ask him to place his anterior teeth edge to edge. Notice that the maxillary central incisor extends distal to the mandibular central incisor because it is the wider tooth.

    The crown is nearly bilaterally symmetrical, and the mesioincisal and distoincisal angles are only a little rounded.

    The mesial and distal surfaces of the crown are nearly straight near the incisal edge; then the crown tapers and becomes narrower from the contact areas to the cervical line.

    The labial surface of the crown is very convex mesiodistally in the cervical third, and nearly flat in the incisal third. (Feel it.)

    Contact areas

    Mesial: In the incisal third near the mesioincisal angle.

    Distal: In the incisal third, about the same level as the mesial contact area.

    Morphology of the labial surface

    The surface of the crown is nearly smooth. There are sometimes 2 very shallow depressions in the incisal third.

Mamelons are usually present on newly emerged teeth. Ordinarily they are soon worn off.

2. ROOT

The root tapers from the cervical line to the apex, and often the apical end is curved distally.

The labial surface of the root is convex mesiodistally.

LINGUAL ASPECT

1. CROWN (Fig. 4-1c)

The crown is narrower on the lingual surface than on the labial.

The lingual surface is smooth and slightly concave in the middle and incisal thirds; the cingulum is convex and small.

The marginal ridges and the lingual fossa usually are scarcely discernible.

2. ROOT

The root is slightly narrower on the lingual side than on the labial.

MESIAL ASPECT

1. CROWN

From this aspect the crown resembles a wedge.

The incisal edge is *lingual* to the axis line of the root; the root apex usually is on the axis line.

The crest of curvature on the labial surface of the crown is just incisal to the cervical line.

The crest of curvature on the lingual surface is on the cingulum.

The labial contour of the crown from the crest of curvature to the incisal edge is so slightly curved that often it seems to be nearly flat. (Feel it.)

The lingual contour is convex over the cingulum, concave in the middle third, and nearly straight in the incisal third.

The mesial surface is nearly flat in the cervical third, nearly flat in the middle third, and convex in the incisal third.

The cervical line has a deep curvature extending incisally about $\frac{1}{3}$ of the crown length. The curvature is deeper on the mesial surface than on the distal.

2. ROOT

The facial and lingual sides of the root are nearly straight from the cervical line to the middle third; then the root tapers. The apex is usually on the axis line. A slight depression is on the mesial surface of the root.

DISTAL ASPECT

1. CROWN

The shape is similar to that seen from the mesial aspect.

The cervical line curves less on the distal surface than on the mesial.

2. ROOT

Usually the depression on the distal surface of the root is more distinct than the one on the mesial surface.

INCISAL ASPECT

To follow this description the tooth should be held in such a position that the incisal edge is toward the observer, the labial surface is at the top, and the observer is looking *exactly* along the root axis line (Fig. 4-1b).

The mandibular central incisor is almost bilaterally symmetrical.

Usually the incisal edge is at right angles to the *labiolingual* root axis plane.

The labiolingual measurement of the crown is *greater* than the mesiodistal measurement. (This is different from the measurements in the maxillary incisors.)

The labial surface is slightly convex in the incisal third.

The lingual surface is slightly concave in the incisal third.

The crown tapers lingually in the middle and cervical thirds: the labial surface is broader than the lingual surface.

Mandibular Lateral Incisor

SIZE (Measured on the labial surface)

Crown length (cervicoincisal measurement)	9.2 mm.
Crown width (greatest mesiodistal measurement)	5.8 mm.
Root length (cervical line to apex)	12.0 mm.

This tooth is a little larger than the mandibular central incisor in the same mouth. The root is often longer.

LABIAL ASPECT

1. CROWN

Shape

The crown of the mandibular lateral incisor resembles that of the mandibular central incisor, but it is not bilaterally symmetrical.

The crown of the lateral incisor is tilted distally on the root, giving the impression that the tooth has been bent at the cervix.

The contact areas of the lateral incisor are not at the same level, a condition different from that found in the central incisor.

Contact areas

Mesial: In the incisal third.

Distal: In the incisal third, but cervical to the level of the mesial contact area.

Morphology of the labial surface

This surface is similar to that of the central incisor.

2. ROOT

The root is tapering, and the apical end often is curved distally.

Its labial surface is convex mesiodistally.

LINGUAL ASPECT

1. CROWN

The shape resembles that of the mandibular central incisor, but lacks bilateral symmetry. The crown is tilted distally on the root.

The cingulum lies slightly distal to the axis line of the root, and the mesial marginal ridge is slightly longer than the distal marginal ridge.

The marginal ridges and the lingual fossa usually are not conspicuous.

2. ROOT

The lingual surface of the root is smooth.

MESIAL ASPECT

1. CROWN

The crown is the shape of a wedge.

The incisal edge is slightly lingual to the axis line of the root.

The labial crest of curvature is just incisal to the cervical line; the lingual crest of curvature is on the cingulum.

The labial surface becomes nearly flat through the middle and incisal thirds of the crown.

The lingual surface is convex over the cingulum, concave in the middle third, and nearly straight in the incisal third.

The mesial surface is nearly flat near the cervical line and in the middle third, and convex in the incisal third. (Feel it.)

The cervical line has a deep curvature extending incisally about $\frac{1}{3}$ of the crown length. The curvature is greater on the mesial surface than on the distal.

2. Root

The root is nearly straight from the cervical line to the middle third; then it tapers. There is a slight longitudinal depression on the mesial surface of the root.

DISTAL ASPECT

1. Crown

The crown is the shape of a wedge.

The cervical line curves less on the distal surface than on the mesial.

2. Root

Usually the depression on the distal surface of the root is more distinct than the depression on the mesial.

INCISAL ASPECT

The mandibular lateral incisor is not bilaterally symmetrical.

Usually the incisal edge does not lie in a straight line mesiodistally, but rather it has a *distolingual twist:* that is, the distal end of the incisal edge is bent lingually, so that the distoincisal angle is more lingual in position than the mesioincisal angle. This twist of the incisal edge corresponds to the curvature of the mandibular dental arch: a tooth on the right side of the arch is twisted clockwise; on the left it is twisted counterclockwise.

The crown is broader labiolingually than mesiodistally.

The labial surface is slightly convex in the incisal third; the lingual surface is slightly concave.

The crown tapers lingually in the middle and cervical thirds.

The cingulum is slightly distal to the axis line of the root.

Variations in Mandibular Incisor Teeth

There is more uniformity of shape in the mandibular incisor teeth than in other teeth.

Occasionally a mandibular lateral incisor is found to have a labial and a lingual root. This is rare.

In some Mongoloid peoples the cingulum of mandibular incisors is characteristically marked by a short deep groove running cervicoincisally. This groove is often a site of dental caries.

Notes

Have you found any mandibular incisors with a vertical groove on the cingulum, either in your examination of mouths or in extracted teeth?

If so, did the groove show evidence of caries?

Have you found, among your extracted teeth, a mandibular incisor with conspicuous mamelons? If you have one, and if you carefully grind off the facial surface in the incisal third of the crown (see *Introduction* for method of grinding) you may find a deep pit between the center mamelon and the mesial and distal mamelons. There may be some evidence of early caries in such a pit.

Do you find any 2-rooted lateral incisors?

5

Permanent Canines

(Cuspids)

General Description of Canines

There are 4 canine teeth: 1 on either side in the maxillary and mandibular arches.

The canine teeth are distal to the lateral incisors, the third teeth from the midline. The mesial surface of the canine is in contact with the distal surface of the lateral incisor.

The canine teeth are sometimes said to be located at the *corners* of the mouth.

DEVELOPMENTAL DATA

	Maxillary	*Mandibular*
Beginning of enamel matrix and dentin formation	4–5 mos.	4–5 mos.
Enamel completed	6–7 mos.	6–7 yrs.
Emergence into oral cavity	11–12 yrs.	9–10 yrs.
Root completed	13–15 yrs.	12–14 yrs.

FUNCTIONS

In dogs, cats, and other animals with long, prominent canine teeth, the function of these teeth is to tear food. In human beings these teeth usually function with the incisors in cutting food.

GENERAL CHARACTERISTICS OF CANINE TEETH (MAXILLARY AND MANDIBULAR)

The canines are the longest teeth in the mouth; they have particularly long roots.

Canine teeth do not ordinarily have mamelons.

The labial surface of a canine tooth is prominently convex.

The incisal edge of a canine tooth is pointed into a cusp rather than nearly straight across as in the incisors. The *mesial slope* of the cusp is shorter than the *distal slope*. (In older individuals the lengths of the cusp slopes are often altered by attrition.)

The distal contact area is more cervical in position than the mesial.

The measurement of the crown is greater labiolingually than it is mesiodistally.

ANTERIOR TEETH

Incisor and canine teeth make up what are known collectively as the *anterior teeth*.

Maxillary Canine

SIZE (Measured on the labial side)

Crown length (cervicoincisal measurement)	10.0 mm.
Crown width (greatest mesiodistal measurement)	7.9 mm.
Root length (cervical line to apex)	17.0 mm.

Notice that the crown length is about the same as that of the maxillary central incisor, but the root is much longer.

LABIAL ASPECT (Examine several extracted teeth as you read. Hold maxillary teeth with crowns down and mandibular teeth with crowns up.)

1. CROWN (Figs. 5-1a and 5-2a)

    Shape

    Instead of a broadly curved incisal edge such as is found in incisor teeth, the incisal edge of a maxillary canine comes to a distinct point. This point is called a *cusp*.

    The cusp has a *mesial slope* and a *distal slope*. (These slopes are sometimes called the mesial and distal incisal ridges.) The mesial slope is *shorter* than the distal.

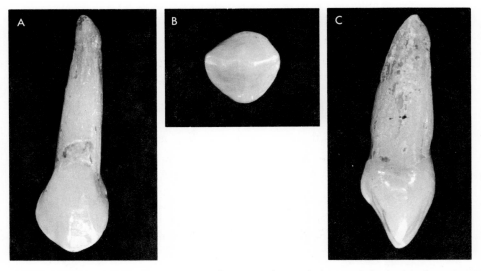

FIG. 5-1a. Maxillary right canine. Labial surface. The scar on the cervical part of the root is damage, possibly due to tooth removal; it is not part of the tooth anatomy. b. Incisal aspect. The labial side is toward the top; mesial at right. c. Distal surface. Notice the curvatures of the facial and lingual surfaces.

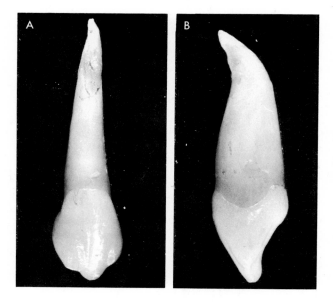

FIG. 5-2a. Maxillary right canine. Labial surface. Both this tooth and the tooth in figure 5-1a are of shapes frequently found. Notice the difference. **b.** Mesial surface of the tooth shown in figure 5-2a. Compare the facial and lingual curvatures with the same curvatures of the tooth in figure 5-1c. The cervical line is clear in this picture. Notice that the enamel fits onto the crown somewhat like a cap, and at the cervical border it appears as a beveled edge. There is *not* a ditch around the tooth at the cementoenamel junction. Examine your extracted teeth carefully and notice the form of the cementoenamel junction.

The *cusp tip* usually is on the root axis line (unless the tooth is worn).

The outline of the mesial side of the crown may be either convex or concave in the cervical third.

The distal side of the crown is slightly concave in the cervical third. It is convex in the middle third.

Contact areas

Mesial: At the junction of the incisal and middle thirds.

Distal: In the middle third. Sometimes just cervical to the junction of the incisal and middle thirds; often in the middle of the middle third.

Morphology of the labial surface

The *labial ridge* runs cervicoincisally near the center of the crown in the middle and incisal thirds. On some teeth this ridge is prominent.

Shallow depressions lie mesial and distal to the labial ridge. (This is said to indicate that the facial side of the crown is made up of three labial lobes. The cingulum on the lingual side of the crown is the fourth lobe.)

2. ROOT

The root is long, slender, and conical.

The labial surface is very convex, and the apical end often is bent distally.

LINGUAL ASPECT

1. CROWN

The crown is narrower on the lingual side than on the labial.

The *cingulum* is large. Its incisal border is sometimes pointed in the center, resembling a small cusp. However, this usually is not called a cusp.

The *mesial* and *distal marginal ridges* usually are of moderate size, and a *lingual ridge* runs cervicoincisally on the center of the lingual surface. It extends from the cingulum to the cusp tip.

The *mesial* and *distal lingual fossae* lie on either side of the lingual ridge and usually are shallow.

Sometimes the lingual surface is smooth, and the ridge and fossae are not discernible.

2. ROOT

The root is narrower on the lingual side than on the labial.

MESIAL ASPECT

1. CROWN (Fig. 5-2b)

The crown is wedge-shaped.

When compared to the incisor teeth, the crown appears thick from the mesial aspect. The cusp tip is usually *labial* to the root axis line—a not unexpected position since this tooth overlaps the mandibular teeth when the jaws are closed. (Examine someone's mouth.)

The crest of curvature of the labial surface is in the cervical third of the crown, but it may not be as close to the cervical line as the corresponding curvature in the incisor teeth. The labial surface is much more convex than in the incisors. (Feel it.)

The crest of curvature on the lingual surface is on the cingulum, which makes up the cervical third of the crown length. The remainder of the lingual surface is straight, or slightly concave in the middle third, and straight or slightly convex in the incisal third.

The cervical line curves incisally. The curvature is greater on the mesial surface than on the distal, and it is less on the canine than on the incisors.

2. ROOT

The mesial surface is broad and has a depression running cervicoapically.

The labial side of the root is slightly convex; the lingual side is more convex.

DISTAL ASPECT

1. CROWN (Fig. 5-1c)

The shape from this aspect is similar to the shape from the mesial aspect.

The cervical line has less curvature than on the mesial surface.

The crown surface is often concave cervical to the contact area.

2. ROOT

A depression runs longitudinally on the root. Usually it is less distinct than the depression on the mesial side.

INCISAL ASPECT (Fig. 5-1b)

To follow this description the tooth should be held in such a position that the incisal edge (cusp tip) is toward the observer, the labial surface is at the top, and the observer is looking *exactly* along the root axis line.

The labiolingual measurement of the crown is greater than the mesiodistal measurement.

The cingulum is large and is about in the center mesiodistally.

The labial surface is very convex and the labial ridge is often prominent.

The cusp tip and cusp slopes lie slightly labial to the mesiodistal axis line of the root.

Mandibular Canine

SIZE (Measured on the labial surface)

Crown length (cervicoincisal measurement)	10.3 mm.
Crown width (greatest mesiodistal measurement)	7.0 mm.
Root length (cervical line to apex)	15.7 mm.

LABIAL ASPECT

1. CROWN (Fig. 5-3a)

Shape

The crown appears long and narrow compared to the crown of the maxillary canine.

The mesial side of the crown is slightly convex, and it is nearly in line with the mesial side of the root. This is usually a conspicuous feature of this tooth.

The distal side of the crown may be slightly concave in the cervical third; it is convex incisal to the cervical third.

The crown is not divided into halves by a coronal extension of the root axis line: there is more of the crown distal to the root line than mesial to it. This makes the crown appear to be tilted distally when the root is held in a vertical position.

The *cusp tip* is usually on the root axis line.

The *mesial slope of the cusp* is shorter than the *distal slope*. (Wear on the incisal edge may alter the length of the cusp slopes.)

Contact areas

Mesial: In the incisal third.

Distal: At the junction of the middle and incisal thirds.

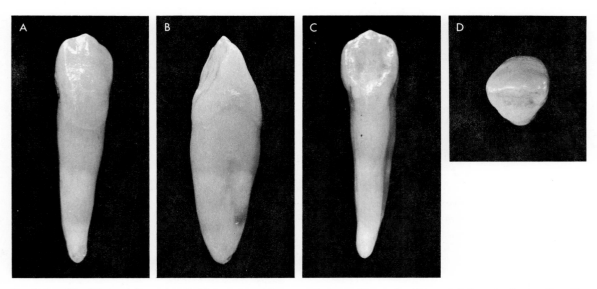

FIG. 5-3a. Mandibular left canine. Labial surface. The cusp tip is a wide angle; the mesial slope is shorter than the distal. **b.** Mesial surface. The crest of curvature of the facial side of the crown is close to the cervical line. **c.** Lingual surface. The curvature of the full length of the tooth that results in a convex line along the mesial side is a frequent characteristic of this tooth. (Of course, there are exceptions.) **d.** Incisal aspect. Labial side at top; mesial at right.

Morphology of the labial surface

The labial surface is smooth and convex.

A labial ridge is often present, but not prominent.

In the incisal third the crown surface is slightly flattened mesial to the labial ridge, and a little more flattened distal to the ridge. (Feel it.)

2. ROOT

The root is convex on the labial surface, and tapers apically.

The apical end is often slightly bent *mesially.*

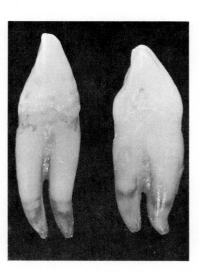

FIG. 5-4. Two mandibular canines, not quite the same shape, but both shapes are frequently found. The furcated root (facial and lingual) is found rarely enough to be interesting, but frequently enough not to be amazing. It would be interesting to learn the frequency of occurrence of this characteristic in different populations.

LINGUAL ASPECT

1. CROWN (Fig. 5-3c)

The crown and the root both taper toward the lingual surface, making it narrower than the labial surface.

The cingulum is low and not prominent. It lies considerably distal to the root axis line.

The marginal ridges are not prominent.

The lingual ridge and the lingual fossae are not prominent.

The mesial marginal ridge is longer than the distal.

2. ROOT

The root is narrower on the lingual side than on the labial.

The lingual surface of the root is convex.

MESIAL ASPECT

1. CROWN (Fig. 5-3b)

The crown is wedge-shaped and thinner in the incisal portion than the crown of the maxillary canine.

The crest of curvature of the labial surface of the crown is close to the cervical line.

The crest of curvature of the lingual surface is on the cingulum. The cingulum is low.

The cusp tip may be on the root axis line, or more often, slightly lingual to the axis line.

The cervical line has a deeper curvature incisally on the mesial surface than on the distal surface.

2. ROOT

The root has a clear depression running longitudinally on the mesial surface.

DISTAL ASPECT

1. CROWN

The general shape from the distal aspect is similar to the shape from the mesial aspect.

The cervical line has less curvature on the distal surface than on the mesial.

The distoincisal angle is slightly more lingual in position than the cusp tip.

2. ROOT

> The longitudinal depression on the distal side of the root is often deeper than the depression on the mesial side.

INCISAL ASPECT (Fig. 5-3d)

The labiolingual measurement of the crown is greater than the mesiodistal measurement.

The cusp tip is near the center labiolingually, or it may be lingual to the center. The distal cusp slope is directed slightly lingually from the cusp tip, placing the distoincisal angle in a position lingual to the position of the cusp tip. This lingual placement of the distoincisal angle gives the incisal part of the crown a *distolingual twist*.

On the labial contour the crest of curvature is mesial to the center line.

On the lingual contour the crest of curvature of the cingulum is distal to the center line.

The crown tapers lingually, making the lingual surface narrower than the labial surface.

Variations in Canine Teeth*

Probably the most conspicuous variation in these teeth is found in the mandibular canine. It is not surprising to find a mandibular canine tooth with the root divided into labial and lingual parts. The division may be only in the apical third or it may extend into the cervical third of the root (Fig. 5-4).

It is rare to find a maxillary canine tooth with the root similarly divided, but this division is known to occur.

*Interesting paper: Taylor, R. M. S., *Variations in Form of Human Teeth: II. An Anthropologic and Forensic Study of Maxillary Canines.* J. Dent. Res., *48*:173–182, 1969.

Notes

Have you seen a maxillary canine with a cusp-like projection on the incisal border of the cingulum?

Have you seen a maxillary canine with a very sharp cusp tip, a prominent labial ridge, and deep depressions mesial and distal to the ridge?

Have you found a mandibular canine with a facial and a lingual root? If so, how far cervically was the root divided?

If you have found a maxillary canine with a root divided into facial and lingual parts, you have a "real find."

6

Permanent Maxillary Premolars

(Bicuspids)

General Description of Maxillary Premolars

There are 4 maxillary premolar teeth: a first and second premolar on either side of the maxillary arch.

The premolars are the fourth and fifth teeth respectively from the center line.

The mesial side of the first premolar is in contact with the distal side of the adjacent canine.

The distal side of the first premolar is in contact with the mesial side of the second premolar.

The distal side of the second premolar is in contact with the mesial side of the maxillary first molar.

In position, the premolar teeth succeed the *primary* first and second *molar* teeth when these primary teeth are shed.

DEVELOPMENTAL DATA

	First premolar	*Second premolar*
Beginning of enamel matrix and dentin formation	1½-1¾ yrs.	2-2¼ yrs.
Enamel completed	5-6 yrs.	6-7 yrs.
Emergence into oral cavity	10-11 yrs.	10-12 yrs.
Root completed	12-13 yrs.	12-14 yrs.

FUNCTIONS

The premolar teeth function with the molar teeth in the mastication of food.

POSTERIOR TEETH

The premolars, along with the molars, are known collectively as the *posterior teeth*.   43

GENERAL CHARACTERISTICS OF MAXILLARY PREMOLAR TEETH

In posterior teeth the facial surface is referred to as the *buccal* surface rather than as the labial surface, since it is next to the *cheek* rather than the lip (L. bucca = cheek).

There are some *similarities* between the maxillary premolars and the anterior teeth:

In the premolars the buccal surface shows evidence of 3 facial, or buccal, lobes, i.e., there is a depression on either side of the center of the facial surface of the crown. The lingual cusp is the fourth lobe.

In the premolars the middle facial lobe is most conspicuous, as is true in the maxillary canines.

The lingual side of the crown and root are narrower than the facial side.

There are some outstanding *differences* between the premolars and the anterior teeth:

The premolars have occlusal surfaces instead of incisal edges.

The crowns and roots are shorter than those of the anterior teeth.

The first maxillary premolar usually has 2 roots: 1 buccal and 1 lingual.

The lingual cusp of the premolar is regarded as an evolutionary enlargement of the cingulum of the anterior teeth.

The marginal ridges of the premolars are on the occlusal surface in a horizontal position rather than on the lingual surface in a nearly vertical position as in the anterior teeth.

From the mesial and distal aspects the crests of curvature of both the labial and lingual surfaces of the crown are more occlusal in position than the corresponding crests of curvature in the anterior teeth.

Maxillary First Premolar

SIZE (Measured on the buccal surface)

Crown length (cervico-occlusal measurement)	8.5 mm.
Crown width (greatest mesiodistal measurement)	7.0 mm.
Root length (cervical line to apex of buccal root)	12.2 mm.

BUCCAL ASPECT (Examine extracted teeth as you read. Hold crown down, root up.)

1. CROWN (Fig. 6-1a)

   Shape

   The crown is broad at the level of the contact areas and more narrow near the cervix.

   The mesial and distal sides of the crown from the contact areas to the cervical line are nearly straight. In some teeth they are concave.

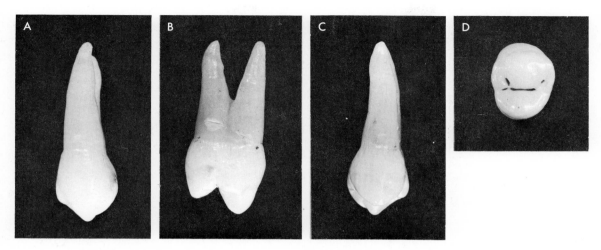

Fig. 6-1a. Maxillary left first premolar. Buccal surface. **b.** Mesial surface. Notice how short the lingual cusp is compared to the buccal. Study the curvatures of the crown outline. **c.** Lingual surface. The buccal cusp is obviously longer than the lingual cusp. **d.** Occlusal surface. Buccal side at top; mesial at left. The mesial marginal groove is clear. The lingual cusp tip is mesial to the center line. The central groove is noticeably longer than that of the maxillary second premolar. See figure 6-5d.

The mesial and distal sides of the crown around the contact areas are convex.

The tip of the buccal cusp often is slightly distal to the vertical axis line of the tooth.

The *mesial slope* of the buccal cusp is *longer* than the distal slope. This is the only tooth, considering canines and the other premolars, in which this is true. Wear, of course, may alter the length of the cusp slopes.

The buccal cusp is relatively long, and in shape resembles the cusp of a maxillary canine.

The mesial and distal slopes of the buccal cusp meet at the cusp tip in an obtuse angle.

Contact areas
    Mesial: Usually in the middle third, just cervical to the junction of the occlusal and middle thirds; but it may sometimes be at the junction of the occlusal and middle thirds.

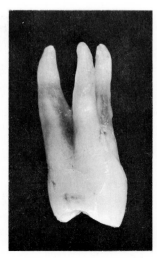

Fig. 6-2. Maxillary left first premolar with 2 buccal roots and 1 lingual root. There is nothing unusual about the crown of this tooth, but the roots are unusual; except for the fact that they are less spread apart, the roots resemble those of a maxillary molar.

Distal: Usually in the middle third; usually slightly more cervical in position than the mesial contact area.

Morphology of the buccal surface

The buccal surface is convex.

There is a prominent elevation running cervico-occlusally in the middle of the crown; it is called the *buccal ridge.*

There is a shallow depression in the occlusal third of the crown mesial and distal to the buccal ridge.

2. ROOT

The roots taper apically and the tip of the buccal root is often slightly bent.

The buccal surface is convex.

LINGUAL ASPECT

1. CROWN (Fig. 6-1c)

The crown is a little narrower on the lingual side than on the buccal.

The lingual cusp is shorter than the buccal cusp.

The mesial and distal slopes of the lingual cusp meet at the cusp tip at a somewhat rounded angle.

The lingual surface of the crown is convex in all areas.

2. ROOT

The lingual surface of the root is convex.

The apical end of the lingual root is sometimes slightly bent, usually distally.

MESIAL ASPECT

1. CROWN (Figs. 6-1b and 6-3)

In an unworn tooth the buccal cusp is about 1 mm. longer than the lingual cusp. This is one of the distinguishing characteristics of the maxillary first premolar.

In an ordinary-sized tooth the distance buccolingually between the cusp tips is about 6 mm. (Since the size of teeth in different individuals is subject to considerable variation, the distance between cusp tips will, of course, vary proportionally.)

Both cusp tips are located within the boundary of the root contour.

Looking at the mesial side of the tooth, the occlusal surface of the crown is seen to slope cervically from each cusp tip to the central groove on the occlusal surface. These slopes are *curved* lines. They are called the *triangular ridges.*

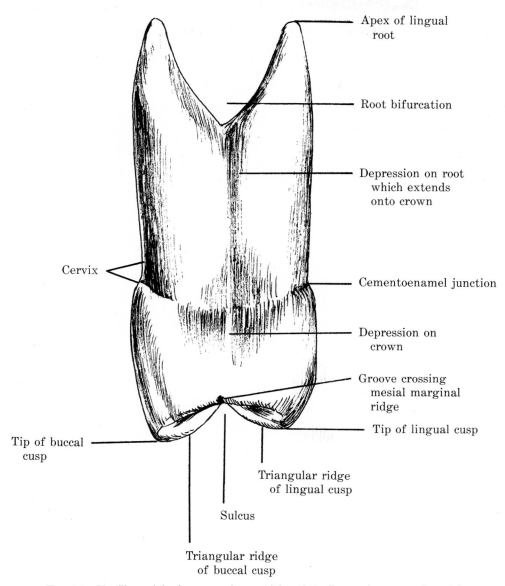

Apex of lingual root

Root bifurcation

Depression on root which extends onto crown

Cervix

Cementoenamel junction

Depression on crown

Groove crossing mesial marginal ridge

Tip of lingual cusp

Tip of buccal cusp

Triangular ridge of lingual cusp

Sulcus

Triangular ridge of buccal cusp

Fig. 6-3. Maxillary right first premolar, mesial surface. Drawn from a tooth model.

The mesial marginal ridge is horizontal and forms the occlusal border of the mesial surface.

The crest of curvature of the buccal surface of the crown is usually at the junction of the middle and cervical thirds.

The crest of curvature of the lingual surface of the crown is usually at the center of the middle third.

The cervical line curves occlusally in a broad curvature. The curvature is greater on the mesial than on the distal side of the tooth.

The depression in the cervical third of the crown is continuous with the depression on the mesial surface of the root. This is one of the distinguishing characteristics of the maxillary first premolar. *This depression is a feature to keep in mind in clinical practice.* (Feel it.)

2. ROOT (Fig. 6-1b)

Often, but not always, there are 2 roots: 1 buccal and 1 lingual. This is a distinguishing characteristic of the maxillary first premolar.

The extent of the bifurcation of the root varies, usually from $\frac{1}{3}$ to $\frac{1}{2}$ of the root length. Sometimes it is nearly the entire length of the root.

Sometimes there is only 1 root. In this case there are usually deep depressions on the mesial and distal root surfaces.

Whether 1 or 2 roots are present, there are 2 root canals.

The depression extending the full length of the undivided cervical part of the mesial surface of the root connects with the depression on the mesial surface of the crown. (Feel it.)

In the 2-root type of maxillary first premolar the buccal and lingual roots are usually relatively straight, except for a curvature which often occurs near the apex.

DISTAL ASPECT

1. CROWN

The distal surface of the crown is similar to the mesial surface, with these differences:

The distal surface is convex: it does not have a depression in the cervical third.
The distal marginal ridge is slightly more cervical in position than the mesial marginal ridge.
The cervical line on the distal surface has less curvature than the cervical line on the mesial surface.

2. ROOT

The distal surface of the root near the cervix is usually convex or flat with little or no depression.

Apical to the convex area, the root has a depression in its undivided portion.

OCCLUSAL ASPECT (Fig. 6-1d)

To follow this description the tooth should be held in such a position that the occlusal surface is toward the observer, the buccal surface is at the top, and the observer is looking *exactly* along the vertical axis.

The buccolingual measurement of the crown is greater than the mesiodistal measurement.

From the occlusal aspect the shape of the *buccal* surface is a wide, inverted V, due to the prominent buccal ridge (Fig. 6-4).

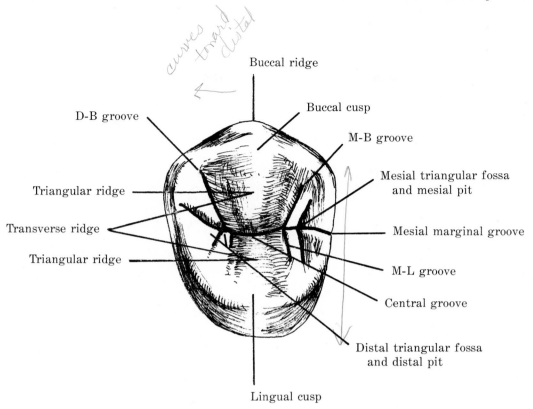

curves toward distal

FIG. 6-4. Maxillary right first premolar, occlusal surface. Drawn from a tooth model.

The lingual side of the tooth is narrower than the buccal side, and the contour is asymmetrical because the lingual crest of curvature is usually mesial to the center line of the tooth.

The tip of the lingual cusp is mesial to the center line of the tooth.

The mesial side of the tooth is nearly straight buccolingually.

The distal side is slightly convex, with a broad curvature around the distal side of the lingual cusp.

Contact areas from the occlusal aspect

Mesial: Near the junction of the buccal and middle thirds. (One-third of the tooth from this aspect means ⅓ of the total buccolingual measurement of the crown, rather than ⅓ of the occlusal surface measurement.)

Distal: Slightly buccal to the mesial contact area.

Morphology of the occlusal surface (Fig. 6-4)

The buccal cusp has a tip, and a mesial and distal cusp slope.

The lingual cusp has a tip, and a mesial and distal cusp slope.

The *buccal triangular ridge* extends from the tip of the buccal cusp lingually across the occlusal surface to the *central groove*.

The *lingual triangular ridge* extends from the tip of the lingual cusp buccally across the occlusal surface and meets the buccal triangular ridge at the central groove.

Combined, the buccal triangular ridge and the lingual triangular ridge form what is known as the *transverse ridge.*

The *central groove* runs mesiodistally across the central part of the occlusal surface. Its length in this tooth is equal to about ½ the mesiodistal width of the occlusal surface. This length is one of the distinguishing characteristics of the maxillary first premolar.

The central groove ends mesially and distally in *mesial* and *distal triangular fossae.*

In the distal triangular fossa the distal end of the central groove meets the *distobuccal* and *distolingual grooves,* and at the point of union there may be a *distal pit.*

In the mesial triangular fossa the mesial end of the central groove meets the *mesiobuccal* and *mesiolingual grooves,* and at the point of union there may be a *mesial pit.*

The *mesial marginal groove,* which crosses the mesial marginal ridge, connects with the mesial end of the central groove in the mesial triangular fossa. The mesial marginal groove is one of the distinguishing characteristics of the maxillary first premolar.

Sometimes there is a groove across the distal marginal ridge also, but usually there is not.

The mesial marginal ridge is relatively straight buccolingually.

The distal marginal ridge is more convex than the mesial marginal ridge.

A NOTE ON PITS AND FISSURES

Associated with grooves are structures of great clinical importance called *pits* and *fissures.*

A *fissure* is a narrow "ditch" which runs along a groove. A groove may or may not have a fissure. A fissure may be relatively shallow, or it may be deep; and it may be so narrow that no dental instrument can be inserted into it.

At the ends of grooves there is often a deep *pit,* which also may be too small to receive even the smallest point of a dental instrument.

Pits and fissures are formed during the development of the tooth; they are *not* the result of damage to the tooth or of dental caries. Because pits and fissures are warm, moist, undisturbed shelters for bacteria, however, they become the site of dental caries more readily than any other area of the tooth.

Both pits and fissures are commonly found in premolar and molar teeth. Pits are found also in certain types of maxillary incisor teeth at the incisal border of the cingulum when there are heavy marginal ridges and a deep lingual fossa. Pits and fissures on all teeth are frequent sites of dental caries.

VARIATIONS

There is considerable variation in maxillary first premolar teeth. Any one or more of the characteristics described may, on some teeth, not appear as here represented (See figure 6-2.).[1]

Maxillary Second Premolar

SIZE (Measured on the buccal surface)

Crown length (cervico-occlusal measurement)	7.8 mm.
Crown width (greatest mesiodistal measurement)	6.8 mm.
Root length (from cervical line to apex)	13.7 mm.

BUCCAL ASPECT

1. CROWN (Fig. 6-5a)

   Shape

   Compared with the first premolars the crown appears less angular and more oblong in shape. (Examine a first premolar and make comparisons.)

   The slopes of the buccal cusp are less steep and the cusp is less pointed than on the first premolar.

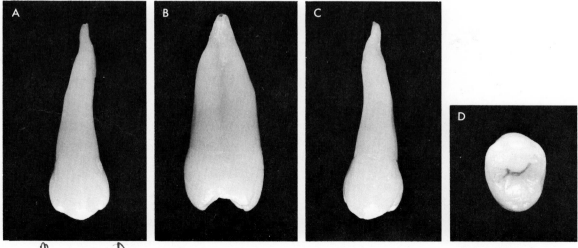

FIG. 6-5a. Maxillary left second premolar. Buccal surface. **b.** Mesial surface. Compare the relative lengths of the buccal and lingual cusps in this tooth with the cusp lengths in the maxillary first premolar. See figure 6-1b. **c.** Lingual surface. **d.** Occlusal surface. The contour from this aspect is nearly bilaterally symmetrical. This is not true of the maxillary first premolar. Notice the length and shape of the central groove compared to that of the first premolar.

The mesial slope of the buccal cusp is *shorter* than the distal slope. (This is different from the first premolar.)

Contact areas

Mesial: Near the junction of the middle and occlusal thirds.

Distal: Usually slightly more cervical in position than the mesial contact area.

Morphology of the buccal surface

The buccal surface is convex and the buccal ridge is less prominent than on the first premolar.

Depressions mesial and distal to the buccal ridge are often difficult to see.

2. ROOT

The root is tapering, and the apical end may be bent distally.

LINGUAL ASPECT

1. CROWN (Fig. 6-5c)

The lingual side is narrower than the buccal and is convex and smooth.

The tip of the lingual cusp is relatively sharp.

The cervical line on the lingual surface is only slightly curved.

2. ROOT

The lingual surface of the root is smooth.

MESIAL ASPECT

1. CROWN (Fig. 6-5b)

The buccal cusp is longer than the lingual cusp, but there is less difference in the length than in the first premolar. (Examine a first premolar and make comparisons.)

The mesial marginal ridge is in a horizontal position and is slightly concave. It is more occlusal in position than the distal marginal ridge.

The crest of curvature on the buccal surface is usually near the junction of the middle and cervical thirds.

The crest of curvature on the lingual surface is about at the center of the middle third of the total crown length.

No depression is present on the mesial surface of the crown.

Sometimes a short groove is close to the mesial marginal ridge, but it usually does not cross the ridge as does the mesial marginal groove on the first premolar.

There is usually a shallow curvature of the cervical line. The curvature is a little greater on the mesial surface than on the distal.

2. ROOT

There is a shallow depression running longitudinally on the mesial surface of the root, but it does not extend onto the crown.

DISTAL ASPECT

1. CROWN

The distal surface of the crown is similar to the mesial surface.

The distal marginal ridge is more cervical in position than the mesial marginal ridge. This is generally true of all posterior teeth, with the exception of the mandibular first premolar.

2. ROOT

The root usually has a depression running longitudinally on the distal surface which is a little deeper than the corresponding depression on the mesial surface. It does not extend onto the crown.

OCCLUSAL ASPECT (Fig. 6-5d)

The occlusal surface of the second premolar is less angular than that of the first premolar. (Compare with a first premolar.)

The buccal ridge on the second premolar is less prominent than on the first premolar.

The crown is often nearly as wide mesiodistally on the lingual side as it is on the buccal side.

The crest of curvature of the lingual side of the crown is near the center line of the tooth or slightly mesial to it. (There is considerable variation in the location of this crest of curvature: it is sometimes located distal to the center line and cannot be relied upon as a criterion for distinguishing a right from a left tooth.)

Seen from the occlusal aspect the second premolar is more nearly bilaterally symmetrical than the first premolar.

The *buccal triangular ridge* and the *lingual triangular ridge* form a *transverse ridge* across the occlusal surface.

The *central groove* runs mesiodistally across the central part of the occlusal surface. In length, it is equal to about ⅓ of the mesiodistal width of the occlusal surface. This is shorter than the central groove of the first premolar.

The central groove ends in the *mesial* and the *distal triangular fossae*.

Contact areas from the occlusal aspect

Mesial: At the junction of the buccal and middle thirds.

Distal: In the middle third; very slightly lingual to the position of the mesial contact area.

Note: The term *premolar* is said to be used to designate any tooth in the *permanent* dentition of mammals which *succeeds* a *primary molar*.

Some dental formulae of permanent dentitions are:

Human	$I\frac{2}{2}$	$C\frac{1}{1}$	$P\frac{2}{2}$	$M\frac{3}{3}$	Dog	$I\frac{3}{3}$	$C\frac{1}{1}$	$P\frac{4}{4}$	$M\frac{2}{3}$
New World Monkey	$I\frac{2}{2}$	$C\frac{1}{1}$	$P\frac{3}{3}$	$M\frac{3}{3}$	Cat	$I\frac{3}{3}$	$C\frac{1}{1}$	$P\frac{3}{2}$	$M\frac{1}{1}$
Old World Monkey	$I\frac{2}{2}$	$C\frac{1}{1}$	$P\frac{2}{2}$	$M\frac{3}{3}$	Cow	$I\frac{0}{3}$	$C\frac{0}{1}$	$P\frac{3}{3}$	$M\frac{3}{3}$
Ape	$I\frac{2}{2}$	$C\frac{1}{1}$	$P\frac{2}{2}$	$M\frac{3}{3}$					

References

1. Morris, D. H., *Maxillary Premolar Variations Among Papago Indians.* J. Dent. Res., *46*:736–738, 1967.

Notes

Have you ground off one side of some single-rooted maxillary first premolars to find out how many root canals there are?

What is the usual importance, clinically, of the depression on the mesial side of the crown of the maxillary first premolar?

What teeth precede the maxillary premolars?

7

Permanent Mandibular Premolars

(Bicuspids)

General Description of Mandibular Premolars

There are 4 mandibular premolar teeth: a first and second premolar on either side of the mandibular arch.

The positions of the mandibular premolars correspond to those of the maxillary premolars: the fourth and fifth teeth from the midline.

The first and second mandibular premolars succeed the first and second *primary* mandibular molars. There are no premolar teeth in the primary dentition.

DEVELOPMENTAL DATA

	First premolar	*Second premolar*
Beginning of enamel matrix and dentin formation	1¾–2 yrs.	2¼–2½ yrs.
Enamel completed	5–6 yrs.	6–7 yrs.
Emergence into oral cavity	10–12 yrs.	11–12 yrs.
Root completed	12–13 yrs.	13–14 yrs.

FUNCTIONS

The mandibular premolars function with the molars in the mastication of food.

GENERAL CHARACTERISTICS OF MANDIBULAR PREMOLARS

There is said to be evidence of the evolutionary development of the mandibular first premolar and of one type of the mandibular second premolar from 3 buccal lobes and 1 lingual lobe. (This is said also to be the case in the incisors, the canines, and the maxillary premolars.)

However, in the case of the second type of mandibular second premolar there is evidence of development from 3 buccal lobes and 2 lingual lobes instead of 1 lingual lobe.

(The custom of referring to the teeth of the human dentition as having "lobes" is not very satisfactory. Such terminology is meaningful only in the study of the probable mode of evolution of teeth from the primitive form to the present complex structures.*)

From the mesial and distal aspects the crowns of the mandibular premolars appear to be tilted lingually at the cervix. This lingual tilting of the crown is characteristic of all mandibular posterior teeth.

From the buccal and lingual aspects many mandibular premolar crowns appear to be tilted distally at the cervix. This distal tilting is characteristic of all mandibular posterior teeth.

The morphologic details of mandibular premolars are difficult to describe because of the great amount of variation in these teeth. To try to list all of the frequent variations would lead to confusion rather than to clarification; therefore, the student must bear in mind while studying these teeth that the description will not fit all teeth despite the plentiful sprinkling of the words *often*, *usually*, and *sometimes* throughout the text.[1,2]

Mandibular First Premolar

SIZE (Measured on the buccal surface)

Crown length (cervico-occlusal measurement)	8.0 mm.
Crown width (greatest mesiodistal measurement)	6.9 mm.
Root length (cervical line to apex)	14.0 mm.

BUCCAL ASPECT (Examine several extracted teeth as you read. Hold crowns up, roots down.)

1. CROWN (Fig. 7-1a)

    Shape

    The mandibular first premolar appears nearly, but not quite, bilaterally symmetrical.

    Sometimes the buccal cusp is located on the center line of the tooth; sometimes it is mesial to the center line.

    The buccal cusp appears sharp, but usually the cusp slopes meet at the cusp tip in an obtuse angle.

    The mesial slope of the buccal cusp is usually shorter than, or the same length as, the distal slope. (Again it must be remembered that wear may alter the length of the cusp slopes. This description is for teeth that are not worn.)

    The crown is narrower mesiodistally in the cervical part than at the contact areas. The mesial and distal sides of the crown from the cervix to the contact areas are nearly straight, or are slightly concave.

    The crown is convex over the contact areas.

*Interesting paper: *Morphology and Evolution of Teeth*. Symposium. J. Dent. Res., *46:5:779–992, 1967.

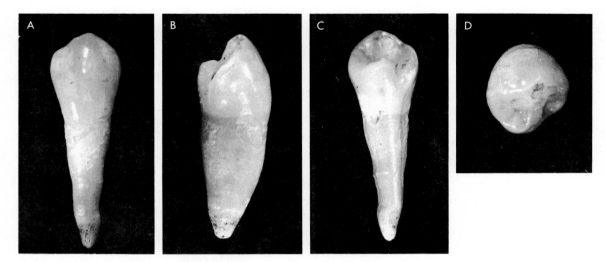

FIG. 7-1a. Mandibular left first premolar. Buccal surface. **b.** Mesial surface. The lingual cusp is very short. The mesiolingual groove lies between the lingual cusp and the mesial marginal ridge. Notice the curvature of the facial and lingual surfaces. The buccal cusp tip is about on the root axis line. **c.** Lingual surface. Much of the occlusal surface is visible because of the shortness of the lingual cusp. In this figure, you can see that the mesial marginal ridge (on the right) is more cervical in position than the distal marginal ridge. **d.** Occlusal surface. The mesial and distal fossae are connected by a barely visible central groove that crosses the transverse ridge.

Contact areas

Mesial: At the junction of the occlusal and middle thirds, or slightly cervical to it.

Distal: The distal contact area is usually on the same level as the mesial contact area. Sometimes it is slightly more *occlusal* in position.

Morphology of the buccal surface

The buccal surface is convex.

The buccal ridge usually is not prominent.

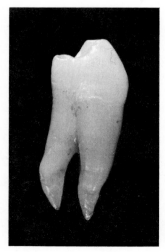

FIG. 7-2. Mandibular left first premolar. Mesial surface. This tooth has a bifurcated root, a condition that seems to be less frequent in the mandibular first premolar than in the mandibular canine. Apparently there are no data on the frequency of occurrence.

Often there is a slight depression in the occlusal third of the buccal surface mesial and distal to the buccal ridge.

2. ROOT

The root is convex on its buccal surface, and is tapering.

It is often bent distally near the tip. (Sometimes it is straight; and it may occasionally be bent mesially.)

LINGUAL ASPECT

1. CROWN (Fig. 7-1c)

The crown is narrower on the lingual side than on the buccal.

The lingual side of the crown is narrower in the cervical third.

The lingual cusp is small; often it is pointed at the tip.

The occlusal surface of this tooth can be seen from the lingual aspect because of the shortness of the lingual cusp.

Usually there is a *mesiolingual groove* separating the mesial marginal ridge from the mesial slope of the lingual cusp. This groove is not always present. (Occasionally a similar groove is present between the distal marginal ridge and the distal slope of the lingual cusp.)

2. ROOT

The root is narrow on the lingual surface, and is tapering.

MESIAL ASPECT

1. CROWN (Fig. 7-1b)

The crown tilts lingually at the cervix.

The tip of the buccal cusp is often centered over the axis line of the root. Sometimes it is a little buccal to the axis.

The tip of the lingual cusp is usually in line vertically with the lingual surface of the cervical portion of the root.

The lingual cusp is shorter than the buccal cusp by about $\frac{1}{3}$ of the total crown length. (The total crown length is measured on the buccal surface.) It is said to be a non-functioning cusp because it is small.

The *triangular ridge of the buccal cusp* slopes cervically from the cusp tip toward the center of the occlusal surface.

The *triangular ridge of the lingual cusp* is often nearly horizontal.

The crest of curvature of the buccal surface of the crown is usually just occlusal to the cervical line. However, in some teeth it is not so close to the cervical line, but still in the cervical third of the crown.

The crest of curvature of the lingual surface of the crown is about in the center of the *total* crown length (*not* the center of the lingual surface of the crown as measured from the cervical line to the tip of the lingual cusp).

Due to the lingual tilting of the crown, the lingual surface of the crown extends lingual to the lingual surface of the root.

The cervical line is curved, with the convexity toward the occlusal surface. It is more occlusal in position on the lingual surface of the tooth than on the buccal surface. In some teeth the difference in the level of the cervical line is as much as 1 mm. from the buccal to the lingual surfaces.

The mesial marginal ridge slopes cervically from the buccal toward the lingual surface.

Usually a *mesiolingual groove* lies between the mesial marginal ridge and the mesial slope of the lingual cusp.

The mesial marginal ridge is usually more cervical in position than is the distal marginal ridge. *This is the only posterior tooth in which this is true.* (It is the relative positions of the mesial and distal marginal ridges on this tooth that sometimes put the mesial contact area cervical to the distal contact area.)

2. ROOT

Often the root is nearly straight on the buccal and lingual sides in the cervical third, or even in the cervical half, and then it tapers apically.

There may be a longitudinal depression in the apical and middle thirds of the mesial surface of the root. This depression may be very deep, or it may be very shallow.

Occasionally the mandibular first premolar will have a furcated root; i.e., the apical part of the root will be divided into a *buccal* and a *lingual* portion (Fig. 7-2).

DISTAL ASPECT

1. CROWN

The general shape resembles that of the mesial aspect.

The distal marginal ridge is less sloped from buccal to lingual than the mesial marginal ridge.

The distal marginal ridge usually is higher above the cervical line than the mesial marginal ridge.

Usually there is no groove between the distal marginal ridge and the distal slope of the lingual cusp. However, in the examination of a large number of teeth, such a groove is sometimes found on both the mesial and the distal sides of the tooth.

The curvature of the cervical line is slightly less on the distal surface than on the mesial.

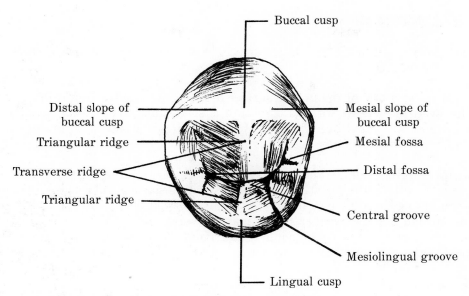

Fig. 7-3. Occlusal surface of a mandibular left first premolar.

2. Root

The root may have a shallow longitudinal depression on the distal surface. This depression usually is not so deep as the one on the mesial side.

3. The Usual Differences Between the Mesial and Distal Aspects Are:

The presence of a mesiolingual groove.
The difference in the direction of slope of the marginal ridges.
The difference in the height of the marginal ridges.
The difference in the depth of the depressions on the root.

OCCLUSAL ASPECT (Figs. 7-1d and 7-3)

To follow this description the tooth should be held with the occlusal surface toward the observer, the buccal surface up, and the observer looking *exactly along the vertical axis of the root.*

There is much variation in the occlusal morphology of the mandibular first premolar teeth.[1] Allowance must be made for variation while following the description.

The tip of the buccal cusp is near the center of the crown outline from the occlusal view.

The cusp slopes of the buccal cusp are in a nearly straight line mesiodistally.

The occlusal surface is broadest mesiodistally at a point just lingual to the line of the buccal cusp slopes and the cusp tip.

The *contact areas as seen from the occlusal aspect* are at the point of broadest mesiodistal dimension, which, *if* the buccal cusp tip is on the axis line, is very slightly lingual to the center of the total crown outline.

The crown converges lingually from the contact areas.

The buccal ridge usually is not prominent.

The crest of curvature of the buccal surface usually is slightly mesial to the center line of the tooth.

The crest of curvature of the lingual surface often is distal to the center line of the tooth. However, the location of this curvature is so variable that it is impossible to make a general statement describing it. A number of teeth should be examined in order that this variability may be seen.

The distal marginal ridge is sometimes nearly at right angles to the buccal surface. However, there is much variation. Sometimes the mesial marginal ridge is more nearly at right angles to the buccal surface than is the distal marginal ridge; and again, the mesial and distal marginal ridges may converge lingually in such a way that the occlusal surface is nearly an equilateral triangle with the base made up of the buccal cusp slopes and the apex of the lingual cusp tip. Compare a number of teeth.

There are a *mesial fossa* and a *distal fossa*.

The *triangular ridge of the buccal cusp* slopes cervically from the cusp tip.

The *triangular ridge of the lingual cusp* is often nearly horizontal.

Sometimes the 2 triangular ridges unite smoothly near the center of the occlusal surface and form a *transverse ridge* which completely separates the mesial and distal fossae. In this case there is sometimes a *mesial groove* extending from the ridge to the mesial fossa and a *distal groove* extending from the ridge to the distal fossae.

Sometimes the 2 triangular ridges form a transverse ridge which is crossed near the center of the occlusal surface by a *central groove* which extends from the mesial to the distal fossa.

Whatever the arrangement of the grooves on the occlusal surface, there may be fissures at the bottom of the grooves. There may also be deep pits in the mesial and distal fossae at the ends of the grooves.
The possibility of there being fissures and pits is, of course, always present wherever there are grooves on teeth. Their chief significance is their susceptibility to caries.

Mandibular Second Premolar

SIZE (Measured on the buccal surface)

Crown length (cervico-occlusal measurement)	8.2 mm.
Crown width (greatest mesiodistal measurement)	7.1 mm.
Root length (cervical line to apex)	14.4 mm.

GENERAL CHARACTERISTICS

There are 2 common types of mandibular second premolars:[2]

   a. The 2-cusp type: 1 buccal cusp and 1 lingual cusp.
   b. The 3-cusp type: 1 buccal cusp and 2 lingual cusps.

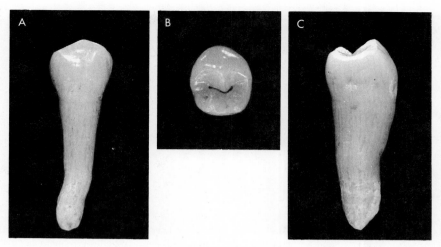

FIG. 7-4a. Mandibular right second premolar. Buccal surface. Three-cusp type. **b.** Occlusal surface (3 cusps). Buccal side at top; mesial at left. The 2 lingual cusps of this tooth are nearly equal in size. Here, it is difficult to tell which is larger; but the mesiolingual cusp (on the left) is larger than the distolingual. Notice the configuration of the central groove. **c.** Distal surface. Notice the difference in cusp height, and also the curvature of the buccal and lingual surfaces.

BUCCAL ASPECT

1. CROWN (both 2-cusp type and 3-cusp type) (Fig. 7-4a)

   Shape

   The shape of the buccal surface is similar in both types.

   The crown bears some resemblance to the mandibular first premolar, but there are differences which make the teeth easily distinguishable. (Examine a first premolar.)

   > The buccal cusp is less pointed than on the mandibular first premolar, and the cusp slopes are less steep.

   > The crown has a somewhat square appearance because it is less constricted at the cervical third than the first premolar.

   Contact areas

   Mesial: Occlusal to the junction of the occlusal and middle thirds.

   Distal: At the same level as the mesial contact area, or slightly cervical to it.

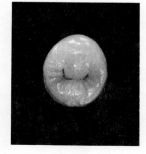

FIG. 7-5. Occlusal surface of a mandibular left second premolar of the 2-cusp type. Buccal side at top; mesial at right. The single lingual cusp is mesial to the center line. The central groove is U-shaped. When compared with the 3-cusp premolar in figure 7-4b, the occlusal surface of this 2-cusp premolar has a round appearance.

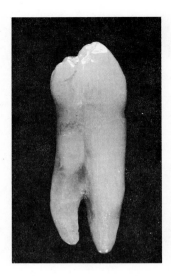

Fig. 7-6. Mandibular right second premolar. Distal surface. This tooth has a buccal and a lingual root. Root furcation seems to be less frequent in this tooth than in the first premolar.

Morphology of the buccal surface

The surface is convex; the buccal ridge is inconspicuous, and sometimes is not discernible.

If the buccal ridge is discernible, there may be either a flattened area or a slight depression on the occlusal third of the buccal surface mesial and distal to the ridge.

2. ROOT (both types of tooth)

The root is slightly wider mesiodistally, and therefore appears heavier than the root of the first premolar.

It is convex on the buccal surface.

It tapers apically and the apical end is often bent distally.

LINGUAL ASPECT

1. CROWN

a. The *2-cusp type* (1 buccal cusp and 1 lingual cusp):

The lingual surface is convex and smooth.

It is narrower than the buccal surface, but not so narrow as the lingual surface of the first premolar.

The lingual cusp is smaller than the buccal cusp, but it is larger than the lingual cusp of the first premolar.

Sometimes the lingual cusp is on the center line of the root; sometimes it is mesial to the center line.

The mesial and distal slopes of the lingual cusp merge into the mesial and distal marginal ridges.

There is sometimes a slight depression where the distal slope joins the distal marginal ridge.

b. The *3-cusp type* (1 buccal cusp and 2 lingual cusps):

The lingual side of the crown is usually narrower than the buccal side, but sometimes it appears nearly the same, or even wider.

The mesiolingual cusp is usually larger than the distolingual cusp. Occasionally they are about equal in size.

There is a groove between the mesiolingual and distolingual cusps. It usually extends slightly onto the lingual surface of the crown.

2. ROOT (both types of tooth)

The root is tapered.

It is slightly narrower on the lingual surface than on the buccal surface.

MESIAL ASPECT

1. CROWN (both types of tooth)

The crown tips lingually.

The crest of curvature of the buccal surface of the crown is usually in the cervical third; but occasionally it is at the junction of the cervical and middle thirds.

The crest of curvature of the lingual surface is about in the middle of the middle third of the total crown length.

The cervical line slopes occlusally: its position on the lingual surface is occlusal to its position on the buccal surface.

In the 2-cusp type, the lingual cusp is shorter than the buccal cusp, but not so much shorter as in the first premolar.

In the 3-cusp type the mesiolingual cusp is shorter than the buccal cusp. The mesiolingual cusp conceals the still shorter distolingual cusp when the tooth is examined from the mesial aspect.

The tip of the buccal cusp is usually located on a line that would divide the crown vertically into buccal and middle thirds.

The tip of the lingual cusp or of the mesiolingual cusp is usually about on a line with the lingual surface of the root.

2. ROOT (both types of tooth)

The root tapers apically.

In some teeth there is a shallow longitudinal depression on the middle and apical thirds of the mesial surface of the root.

Occasionally, but not often, the root is furcated (Fig. 7-6).

DISTAL ASPECT (Fig. 7-4c)

1. CROWN (both types of tooth)

The shape of the distal aspect of the crown resembles the shape of the mesial aspect.

The distal marginal ridge is slightly concave and is more cervical in position than the mesial marginal ridge. This is true of all mandibular posterior teeth except the mandibular first premolar.

The cervical line is less curved on the distal surface than on the mesial surface.

The distolingual cusp is usually smaller than the mesiolingual cusp.

2. ROOT

The distal surface of the root often has a shallow longitudinal depression in the middle and apical thirds.

OCCLUSAL ASPECT

To follow this description the tooth should be held in the same position as that used in examining the first premolar.

a. *The 2-cusp type* (Fig. 7-5):

From the occlusal aspect the crown appears to be a little narrower and a little more rounded on the lingual side than on the buccal side.

The lingual cusp is smaller than the buccal cusp.

There is, of course, a triangular ridge on both the buccal and lingual cusps.

The tip of the buccal cusp is usually at the junction of the buccal and middle thirds of the occlusal aspect. (Recall that in the description of the mesial aspect it was said that the tip of the buccal cusp was on a line dividing the buccal from the middle third of the crown.)

There is a *central groove* extending mesiodistally across the occlusal surface. Sometimes this groove is short and nearly straight; sometimes it is U-shaped with the open end of the U directed buccally; sometimes it is interrupted near its center by a union of the buccal and lingual triangular ridges.

The central groove ends in the *mesial* and *distal fossae*, where it often joins a *mesiobuccal* and a *distobuccal groove*.

There may be a depression separating the distal marginal ridge from the distal slope of the lingual cusp.

b. *The 3-cusp type* (Figs. 7-4b and 7-7):

The occlusal surface is more nearly square than is the occlusal surface of the 2-cusp type because the crown is broad on the lingual side. Sometimes, if the lingual cusps are large, the occlusal surface is broader on the lingual side than on the buccal side.

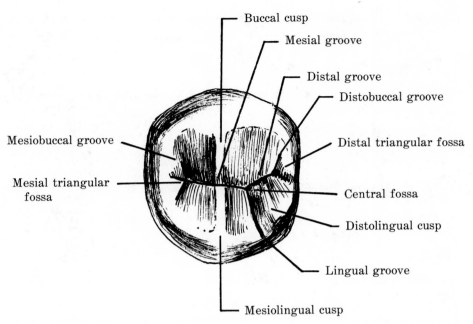

Buccal cusp

Mesial groove

Distal groove

Distobuccal groove

Mesiobuccal groove

Distal triangular fossa

Mesial triangular fossa

Central fossa

Distolingual cusp

Lingual groove

Mesiolingual cusp

Fɪɢ. 7-7. Mandibular right second premolar (3-cusp type). Occlusal surface. Drawn from a tooth model.

The mesiolingual cusp is usually larger than the distolingual. The difference in size may be little or great. Sometimes the cusps seem to be nearly equal in size.

The *lingual groove* is located between the 2 lingual cusps.

There is a triangular ridge on both of the lingual cusps and on the buccal cusp.

The triangular ridges converge toward the central fossa.

The *central fossa* lies about midway between the buccal cusp tip and the lingual border of the occlusal surface, and slightly distal to the center of the occlusal surface.

The *mesial groove* extends from the central fossa to the *mesial triangular fossa*.

The *distal groove* extends from the central fossa to the *distal triangular fossa*. It is shorter than the mesial groove.

The *lingual groove* extends from the central fossa lingually between the mesiolingual and the distolingual cusps and onto the lingual surface of the crown.

HOW TO DISTINGUISH THE MESIAL SIDE FROM THE DISTAL SIDE OF THESE TEETH

a. *The 2-cusp type:*

This is often difficult due to the variability of this tooth.

There is often a distolingual groove separating the distal marginal ridge from the lingual cusp.

The distal marginal ridge usually is a little more cervical in position than the mesial marginal ridge.

The tip of the root often is bent distally.

The cervical line is more curved on the mesial surface than on the distal surface.

b. *The 3-cusp type:*

The distolingual cusp is often smaller than the mesiolingual cusp.

The distal groove is shorter than the mesial groove.

The central fossa is distal to the center of the occlusal surface.

GENERALITIES

Several characteristics found in mandibular premolars are found also in all other mandibular posterior teeth:

The crowns appear to be tilted lingually.

The crowns usually tip distally. (Sometimes mandibular *second* premolars do not have this distal tilt to the crown.)

The distal marginal ridge is more cervical in position than the mesial marginal ridge on all mandibular posterior teeth *except* the mandibular first premolar.

In the mandibular first premolar (the exception) the mesial marginal ridge is more cervical in position than the distal marginal ridge.

References

1. Kraus, B. S., and Furr, M. L. *Lower First Premolars, Part I. A Definition and Classification of Discrete Morphologic Traits.* J. Dent. Res., *32:*554, 1953.
2. Ludwig, F. J., *The Mandibular Second Premolar: Morphologic Variations and Inheritance.* J. Dent. Res., *36:*263–273, 1957.

Notes

If you have a large collection of extracted teeth available, you might find it interesting to determine the ratio of 3-cusp to 2-cusp mandibular second premolars. (Remember that your sample is probably too small to make a general conclusion. You know only what you have in your somewhat restricted sampling.)

Examine the mouth of a number of your associates to see if you find any with a 2-cusp second premolar on 1 side of the mouth and a 3-cusp premolar on the other.

In examining mouths, look carefully for premolar teeth that do not correspond to the descriptions given here. You may be surprised.

8

Permanent Maxillary Molars

General Description of Maxillary Molars

There are 6 permanent maxillary molar teeth: a first, second, and third molar on either side of the maxillary arch; the sixth, seventh and eighth teeth from the midline.

The permanent molar teeth have no predecessors in the primary dentition. The *first permanent* maxillary molar emerges into the mouth distal to the *second primary* molar at about the age of 6 years while the second primary molar is still present. When the primary molars are replaced by the premolar teeth of the permanent dentition the second premolar is in contact distally with the mesial surface of the first permanent molar.

The second permanent molar emerges distal to the first at about the age of 12 or 13 years. The third molar emerges distal to the second at 17 to 21 years.

DEVELOPMENTAL DATA

	First molar	*Second molar*	*Third molar*
Beginning of enamel matrix and dentin formation	At birth	2½–3 yrs.	7–9 yrs.
Enamel completed	2½–3 yrs.	7–8 yrs.	12–16 yrs.
Emergence into oral cavity	6–7 yrs.	12–13 yrs.	17–21 yrs.
Root completed	9–10 yrs.	14–16 yrs.	18–25 yrs.

FUNCTIONS

The molar teeth are used in breaking and chewing food.

GENERAL CHARACTERISTICS OF MAXILLARY MOLAR TEETH

These teeth are larger than the other maxillary teeth.

They have 3 roots: mesiobuccal, distobuccal, and lingual (palatal). The lingual root is the longest, the distobuccal root the shortest.

They have 3 or more cusps.

The crowns are much larger both buccolingually and mesiodistally than the crowns of the premolars.

Maxillary First Molar

GENERAL INFORMATION

The first molar teeth are located near the center of the arch anteroposteriorly in the adult.

The first molars emerge into the oral cavity distal to the *primary* second molars while the primary dentition is still intact. This is important to remember because these permanent first molars are sometimes mistaken by parents for primary teeth, and are neglected if the parents happen to be unaware that the primary teeth as well as the permanent teeth need dental care.

The permanent first molars are often the first permanent teeth to become carious; and they are often the first teeth of the permanent dentition to be lost.

SIZE

Crown length (measured on the buccal surface from the tip of the
mesiobuccal cusp to the cervical line)   8.0 mm.
Crown width (measured parallel to the buccal surface from the crest
of curvature on the mesial surface to the crest of curvature on
the distal surface)   11.0 mm.
Root length
Mesiobuccal root (measured on the buccal surface from the cervical
line to the apex)   12.7 mm.
Lingual root (measured on the lingual surface from the cervical
line to the apex)   14.5 mm.

BUCCAL ASPECT (Examine several extracted teeth as you read. Hold roots up, crowns down.)

1. CROWN (Figs. 8-1a and 8-2a)

   Shape

   From the buccal aspect the crown is broad near the junction of the occlusal and middle thirds and narrower near the cervical line.

   There are two cusps on the buccal side: a *mesiobuccal cusp* and a *distobuccal cusp.* The mesiobuccal cusp is the longer and the wider of the 2 cusps.

   Each cusp has a tip, a mesial slope, and a distal slope.

   The *buccal groove* lies between the cusps and extends cervically on the buccal surface to the middle third of the crown. At the end of the buccal groove there is sometimes a pit which may become the site of dental caries.

   From the buccal aspect the distal side of the crown is concave in the cervical third. Occlusal to this concavity the surface is convex.

   The distal part of the buccal surface of the crown is often depressed in the cervical third.

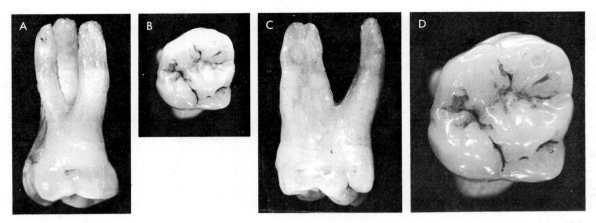

F<small>IG</small>. 8-1a. Maxillary right first molar. Buccal surface. **b.** Occlusal surface. Buccal side at top; mesial at right. This specimen has an unusually large cusp of Carabelli. **c.** Mesial surface. The wide mesiobuccal root hides the narrower distobuccal root. **d.** An enlargement of figure 8-1b. Compare the surface morphology with the labeled diagram in figure 8-3 and with the photograph of the occlusal surface of another maxillary right first molar in figure 8-2d. The groove at the place of attachment of the cusp of Carabelli to the mesiolingual cusp, such as is seen here, is sometimes the site of caries.

Contact areas

Mesial: At the junction of the occlusal and the middle thirds.

Distal: At the middle of the middle third.

2. R<small>OOTS</small>

At the cervical line the crown is attached to the *root trunk*. This is the undivided part of the root.

Apical to the root trunk the root is divided into 3 parts: the *mesiobuccal root*, the *distobuccal root*, and the *lingual root*. The point of furcation is often about at the junction of the cervical and middle thirds of the mesiobuccal root.*

The mesiobuccal and distobuccal roots are often well separated. (This is in contrast to the roots of the second molar, which are often rather close together, and to those of the third molar, which are often fused.)

The mesiobuccal and distobuccal roots usually are curved distally. In the first molar the curvature of the mesiobuccal root may be enough to place the apex of this root distal to the line of the buccal groove on the crown, but there is much variation in the shapes of the roots.

Both the mesiobuccal and distobuccal roots taper apically.

---

*In Mongoloid peoples the maxillary first molars often have a very long root trunk; sometimes there is no furcation at all.[7]

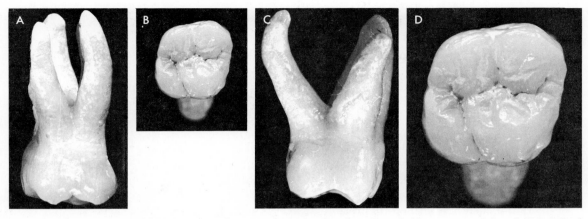

FIG. 8-2a. Maxillary right first molar. Buccal surface. **b.** Occlusal surface. Buccal side at top; mesial at right. **c.** Distal surface. **d.** An enlargement of figure 8-2b. The somewhat wrinkled enamel surface and the relatively shallow fossae and grooves are more easily seen in the enlargement. Notice the square, or nearly square, appearance of this tooth. The cusp of Carabelli is small.

LINGUAL ASPECT

1. CROWN

The crown of the maxillary first molar is often broader mesiodistally on the lingual surface than on the buccal surface, excepting in the cervical third. The cervical third is narrower on the lingual surface.

There are 2 large cusps, the *mesiolingual cusp* and the *distolingual cusp*. The mesiolingual cusp is the larger of the two.

Usually there is a small *fifth cusp* attached to the lingual surface of the mesiolingual cusp. The tip of the fifth cusp is usually a millimeter or more cervical to the tip of the mesiolingual cusp. The fifth cusp varies greatly in shape and size. It may be a conspicuous, well-formed cusp; or, at the other extreme, it may be barely discernible, absent, or there may even be a depression in this location. It is also called the *cusp of Carabelli* or the *tubercle of Carabelli* after the Austrian dentist who described it in 1842. The presence or absence of the cusp of Carabelli seems to be a racial characteristic.

A number of studies have been made concerning the occurrence and size of the cusp of Carabelli.[1,3,4,5,6] One investigator reports that it is extremely rare in the East Greenland Eskimo. In European peoples it is usually present. It is found occasionally on second and third maxillary molars as well as on first molars. Infrequently a fifth cusp may be present on the first molar on one side of the mouth and absent on the other side.

Coming from between the mesiolingual and distolingual cusps is a groove which extends onto the lingual surface. Here it is called the *lingual groove*.

This lingual groove usually is continuous with the longitudinal depression on the lingual surface of the lingual root. It is also continuous with the distal oblique groove on the occlusal surface of the tooth.

2. ROOT

The lingual root is the longest of the roots. It tapers apically.

It often is not curved distally.

There is often a longitudinal depression on the lingual side of the lingual root.

MESIAL ASPECT

1. CROWN (Fig. 8-1c)

The crown measurement cervico-occlusally on the mesiobuccal cusp is usually less than the buccolingual crown measurement. This makes the crown appear short and broad.

Two cusps are seen: the *mesiobuccal cusp* and the *mesiolingual cusp*. The mesiolingual cusp is the longer. The cusps on the distal side of the tooth are shorter, and so usually are not seen when the tooth is examined from the mesial side.

If the cusp of Carabelli is present, it also is seen from the mesial aspect.

The crest of curvature of the buccal side of the crown is usually just occlusal to the cervical line.

The crest of curvature of the lingual side of the crown is in either the cervical third or the middle third, or at the junction of the cervical and middle thirds. Usually in teeth in which the fifth cusp is large the crest of curvature of the lingual side is in the middle third.

The *mesial marginal ridge* connects the *mesiobuccal cusp* and the *mesiolingual cusp*. The ridge is concave.

There is often a mesial marginal groove crossing the mesial marginal ridge.

The cervical line has only a little curvature.

There is sometimes a broad, shallow depression in the center of the cervical third of the mesial surface of the crown. Sometimes this surface is slightly convex.

2. ROOT

The mesiobuccal root is shorter than the lingual root.

Often the apex of the mesiobuccal root is in line with the tip of the mesiobuccal cusp.

The buccal surface of the mesiobuccal root is slightly convex. It often extends a little buccal to the crown.

The lingual surface of the mesiobuccal root is often more convex and, in the apical third, curves sharply toward the apex.

The mesial surface of the mesiobuccal root has a longitudinal depression.

The lingual root is the longest of the 3 roots.

The lingual root tapers apically and usually is curved buccolingually, with the concavity on its buccal surface.

On the first molar the lingual surface of the lingual root usually extends conspicuously lingual to the lingual surface of the crown.

DISTAL ASPECT

1. CROWN (Fig. 8-2c)

From this aspect both the lingual surface and the buccal surface of the crown can be seen because the crown is narrower on the distal side than on the mesial side.

There are 4 cusps clearly visible: the *distobuccal cusp*, the *distolingual cusp*, part of the *mesiobuccal cusp*, and part of the *mesiolingual cusp*. Also, the fifth cusp, which is on the lingual surface of the mesiolingual cusp, can usually be seen.

The distobuccal cusp is often slightly longer than the distolingual cusp.

The cusps of this tooth in the usual order of their height, largest to smallest, are: mesiolingual, mesiobuccal, distobuccal, distolingual, fifth cusp.

The distal marginal ridge is shorter and more concave than the mesial marginal ridge. There is sometimes a distal marginal groove crossing the distal marginal ridge.

The distal surface of the crown usually has a concavity in the cervical third.

There is less curvature to the cervical line on the distal surface than on the mesial surface.

2. ROOTS

The distobuccal root is shorter and more narrow buccolingually than the mesiobuccal root, and it does not extend so far buccally.

The distal surface of the distobuccal root is convex, usually without a longitudinal depression.

OCCLUSAL ASPECT (Figs. 8-1b and d, 8-2b and d, 8-3)

To follow this description the tooth should be held in such a position that the observer is looking exactly perpendicular to the plane of the occlusal surface, with a line visualized through the tips of the buccal cusps in an exactly horizontal position. Because of the spread of the roots, possibly some of each of the 3 roots will be visible when the tooth is in this position.

In studying this tooth it must be remembered that the morphology is extremely variable.

The contour of the occlusal surface is not square, but it gives the general impression of squareness when compared to other teeth. Actually, it is roughly a parallelogram, with 2 acute and 2 obtuse angles.

The acute angles are the mesiobuccal and the distolingual; the obtuse angles are the distobuccal and the mesiolingual.

In some teeth the lingual side of the crown seen from the occlusal aspect is slightly wider mesiodistally than the buccal side.

The distal side is often narrower buccolingually than the mesial side.

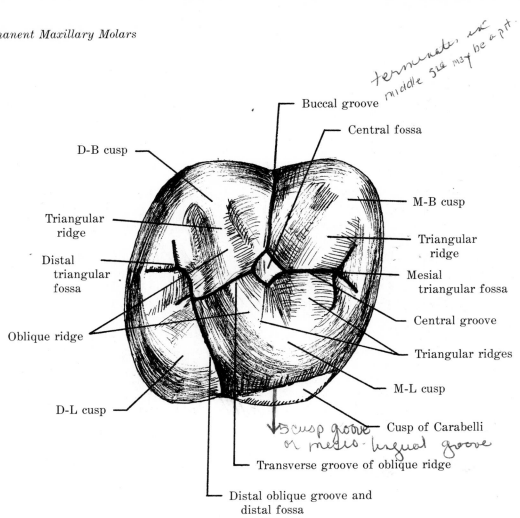

FIG. 8-3.   Occlusal surface of a maxillary right first molar. Drawn from a tooth model.

Assuming that the cusp of Carabelli is present, the 5 cusps are the mesiobuccal, the distobuccal, the mesiolingual, the distolingual, and the fifth cusp. The mesiolingual cusp is conspicuously the largest.

There are ordinarily 4 fossae on the occlusal surface:

> The *central fossa* is near the center of the occlusal surface.
> The *mesial triangular fossa* is just within the mesial marginal ridge.
> The *distal triangular fossa* is just within the distal marginal ridge.
> The *distal fossa* is the elongated fossa between the mesiolingual and the distolingual cusps.

Each of the 4 large cusps has a *triangular ridge*. The triangular ridges of the mesiolingual cusp and the distobuccal cusp meet and form a diagonal ridge called the *oblique ridge*.

The mesiolingual cusp often has a second triangular ridge mesial to the one that meets the triangular ridge of the distobuccal cusp. This second triangular ridge of the mesiolingual cusp meets the triangular ridge of the mesiobuccal cusp and the two together make up a *transverse ridge*.

The *buccal groove of the central fossa* extends buccally from the central fossa and continues onto the buccal surface of the crown as the *buccal groove*.

The *central groove of the central fossa* extends mesially from the central fossa over the transverse ridge and ends in the mesial triangular fossa.

The *transverse groove of the oblique ridge* extends from the central fossa across the oblique ridge to the distal triangular fossa.

The *distal oblique groove* extends from the distal triangular fossa lingually between the distolingual cusp and the mesiolingual cusp (along the distal fossa) and continues onto the lingual surface as the *lingual groove.*

The *fifth cusp groove* separates the fifth cusp from the mesiolingual cusp.

Maxillary Second Molar

GENERAL INFORMATION

The second molar teeth are distal to the first molars. The mesial surface of the second molar is in contact with the distal surface of the first molar; and the distal surface of the second molar is in contact with the mesial surface of the third molar. There is much variation in the morphology of these teeth.

SIZE

Crown length (measured from tip of mesiobuccal cusp to cervical line)	7.5 mm.
Crown width (measured from mesial to distal contact area)	10.2 mm.
Root length	
Mesiobuccal root (measured on buccal surface from cervical line to apex)	12.0 mm.
Lingual root (measured on lingual surface from cervical line to apex)	13.0 mm.

BUCCAL ASPECT

1. CROWN (Fig. 8-4a)

   Shape

   The crown is broader at the junction of the occlusal and middle thirds than in the cervical third, but it usually appears less broad than the crown of the first molar.

   The *mesiobuccal cusp* is larger and longer than the *distobuccal cusp.*

   The crown is tilted distally at the cervix. The occlusal surface slants cervically from mesial to distal.

   A *buccal groove* separates the buccal cusps. It is shorter than the buccal groove of the first molar, and it ends less often in a buccal pit on the buccal surface.

   Contact areas

   Mesial: At the junction of the occlusal and middle thirds.

   Distal: At the middle of the middle third.

2. ROOTS

The roots are sometimes shorter than those of the first molar.

The roots are ordinarily less spread apart than those of the first molar. The *mesiobuccal* and *distobuccal roots* may be less curved than those of the first molar; they may be nearly parallel with each other. (There is much variation.)

The root trunk resembles that of the first molar.

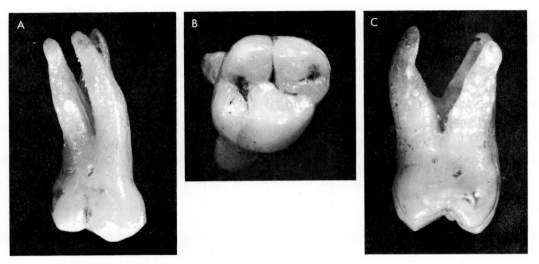

FIG. 8-4a. Maxillary right second molar. Buccal surface. The roots are less spread than those of the first molar. **b.** Occlusal surface. Buccal side at top; mesial at right. The mesiobuccal angle is smaller and the mesiolingual angle larger than in the first molar. Sometimes this difference between the shape of the first and second molars is greater than is seen here. Often in the second molar the acute angles are smaller and the obtuse angles larger than in this tooth; and often the distolingual cusp is much smaller, or even missing. **c.** Distal surface.

LINGUAL ASPECT

1. CROWN

The crown usually appears smaller than that of the first molar in the same mouth.

The cervical third of the crown is narrower than the middle third.

There are visible a *mesiolingual cusp* and a *distolingual cusp*. The mesiolingual cusp is the larger.

There is a *second type of maxillary second molar* in which the distolingual cusp is absent, leaving just 1 lingual cusp, which is large.

The lingual groove separates the mesiolingual and the distolingual cusps. (This groove is, of course, absent if there is no distolingual cusp.)

Usually there is no fifth cusp on the second maxillary molar, but in some individuals a fifth cusp is present on these teeth.

2. ROOTS

The lingual root resembles the lingual root of the first molar.

MESIAL ASPECT

The crown looks very much like that of the first molar except that usually there is no fifth cusp.

The roots are less spread apart than the roots of the first molar. The lingual root usually does not extend beyond the lingual surface of the crown.

The mesiobuccal root usually does not extend buccally beyond the buccal surface of the crown.

DISTAL ASPECT (Fig. 8-4c)

From the distal side much of the buccal surface and much of the lingual surface of the crown can be seen because the crown is narrower on the distal side than on the mesial.

The apex of the lingual root is often in line with the tip of the distolingual cusp.

The mesiobuccal root, which is larger than the distobuccal root, can be seen behind the distobuccal root.

OCCLUSAL ASPECT (Fig. 8-4b)

Compared with the first maxillary molar:

The crown is narrower on the lingual side than on the buccal side. This often is not true of the first molar.

The crown is narrower on the distal side than on the mesial. This is usually more pronounced in the second molar than in the first.

Usually there is no fifth cusp.

The distolingual cusp is proportionally smaller in the second molar. Sometimes it is absent so that the tooth has only 3 cusps.

The 4-cusp type of second molar is less square in appearance than the first molar: the angles at the distobuccal and the mesiolingual cusps are wider; and the angles at the mesiobuccal and distolingual cusps are smaller.

The occlusal shape of the 3-cusp type of second molar (no distolingual cusps) is somewhat triangular, the apex of the triangle being the lingual cusp.

The following structures are similar to those of the first molar:

Cusps
Mesiobuccal, distobuccal
Mesiolingual, distolingual (sometimes absent)

Fossae
Central fossa
Mesial triangular fossa
Distal triangular fossa
Distal fossa (if distolingual cusp is present)

Ridges
    Mesial and distal marginal ridges
    Triangular ridges
    Oblique ridge
    Transverse ridge

Grooves (principal)
    Central groove of central fossa
    Buccal groove of central fossa
    Transverse groove of oblique ridge
    Distal oblique groove (if distolingual cusp is present)

Maxillary Third Molar

GENERAL INFORMATION

The maxillary third molar is in contact mesially with the distal surface of the second molar.

The distal surface of the third molar is not in contact with any tooth.

SIZE

The size is variable. The third molar is ordinarily smaller than the first or the second molar.[*] The roots are ordinarily shorter.

DESCRIPTION (Fig. 8-5)

1. CROWN

The great amount of variation in the third molars makes a general description difficult.

Occasionally the crown of the third molar resembles that of a first molar; occasionally it resembles that of a second molar.

[*] In a study of relative tooth size[2] the second maxillary molar was reported to be larger than the first maxillary molar in 33 per cent of a sample of Ohio Caucasian population, and in 36 per cent of a Pima Indian population. In the mandibular arch this size sequence was seen in 10 per cent of Ohio Caucasians and in 19 per cent of Pima Indians.

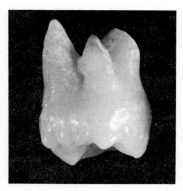

FIG. 8-5. Maxillary right third molar. Distal surface. Third molars do not always have short roots, and they do not always have fused roots. Occasionally their morphology is indistinguishable from that of some second molars.

Nearly always when the crown resembles that of a first or second molar it can be distinguished from these teeth because of the presence of numerous supplemental grooves, particularly on the occlusal surface. These supplemental grooves often give the occlusal surface a wrinkled appearance.

If a distolingual cusp is present it is usually small.

Sometimes the form of the crown is so irregular that it is difficult to identify the mesiobuccal, the distobuccal, and the lingual cusps.

2. Root

There are 3 roots: mesiobuccal, distobuccal, and lingual.

Sometimes the 3 roots are separated, as in the first and second molars. Sometimes the furcation extends only a short distance cervically from the apices of the roots, so that the trunk is long. Sometimes the roots are fused from the cervix to the apices.

Usually the roots are shorter than on the first and second molars, and often they are very crooked.

References

1. Carbonelli, V. M., *The Tubercle of Carabelli in the Kish Dentition, Mesopotamia, 3000 b.c.* J. Dent. Res., *39:*124, 1960.
2. Garn, S. M., Lewis, A. B., and Kerewsky, R. S., *Molar Size Sequence and Fossil Taxonomy.* Science, *142:*1060, 1963.
3. Garn, S. M., Lewis, A. B., Kerewsky, R. S., and Dahlberg, A. A., *Genetic Independence of Carabelli's Trait from Tooth Size or Crown Morphology.* Arch. Oral Biol., *11:*745-747, 1966.
4. Kraus, B. S., *Carabelli's Anomaly of the Maxillary Molar Teeth. Observations on Mexican and Papago Indians and an Interpretation of the Inheritance.* Amer. J. Hum. Genet., *3:*348, 1951.
5. Kraus, B. S., *Occurrence of the Carabelli Trait in the Southwest Ethnic Groups.* Amer. J. Phys. Anthrop., *17:*117, 1959.
6. Meredith, H. V., and Hixon, E. H., *Frequency, Size, and Bilateralism of Carabelli's Tubercle.* J. Dent. Res., *33:*435, 1954.
7. Tratman, E. K., *A Comparison of the Teeth of People. Indo-European Racial Stock and Mongoloid Racial Stock.* The Dental Record, *Vol. LXX.* No. 2., pp. 31-53, Feb. 1950; and No. 3, pp. 63-88, Mar. 1950.

Notes

Examine the teeth of your associates and notice the variation in the cusp of Carabelli. Many people have been intrigued by this little cusp. (See references on p. 81 for papers that carry extensive bibliographies.)

How many second molars do you find with only 1 lingual cusp (distolingual cusp absent)?

How many molars do you find in your collection of extracted teeth that you are unable to identify as first, second, or third with reasonable assurance? (If you feel *no* uncertainty about *any* teeth in a fairly large collection maybe you need to study more.)

9

Permanent Mandibular Molars

General Description of Mandibular Molars

There are 6 permanent mandibular molar teeth: a first, second, and third molar on either side of the mandibular arch.

The mandibular molars are the sixth, seventh, and eighth teeth from the midline.

In the adult dentition the mandibular first molar is distal to the mandibular second premolar. It emerges distal to the *primary* second molar at about 6 years of age while the primary teeth are still present. *The permanent mandibular molars have no predecessors in the primary dentition.*

In the adult the permanent first molar is about in the center of the mandibular arch, anteroposteriorly.

The second molar is distal to the first and is in contact with the distal surface of the first molar and the mesial surface of the third molar.

The third molar is the last tooth in the arch and its distal surface is not in contact with any other tooth.

DEVELOPMENTAL DATA

	First molar	Second molar	Third molar
Beginning of enamel matrix and dentin formation	At birth	2½–3 yrs.	8–10 yrs.
Enamel completed	2½–3 yrs.	7–8 yrs.	12–16 yrs.
Emergence into oral cavity	6–7 yrs.	11–13 yrs.	17–21 yrs.
Roots completed	9–10 yrs.	14–15 yrs.	18–25 yrs.

FUNCTIONS

Crushing and chewing food.

GENERAL CHARACTERISTICS OF MANDIBULAR MOLAR TEETH

The crowns of mandibular molars are oblong in shape: they are larger mesiodistally than buccolingually and larger mesiodistally than cervico-occlusally.

The crowns taper lingually and distally so that the lingual side is narrower than the buccal and the distal side is narrower than the mesial.

The crown tilts distally; the occlusal surface slopes toward the cervix from mesial to distal.

When examined from the mesial or distal sides, the crown appears to be tilted lingually.

There are 2 roots: a mesial root and a distal root.[10]

The root furcation usually is close to the cervical line; the root trunk usually is short.

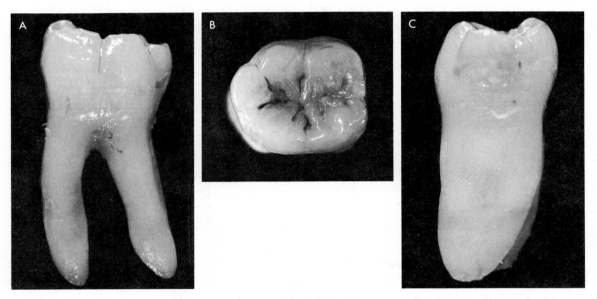

FIG. 9-1a. Mandibular left first molar. Buccal surface. **b.** Occlusal surface. Buccal side at top; mesial at right. Compare this with the drawing in figure 9-2. **c.** Distal surface. Buccal side at left. Notice the curvatures of the buccal and lingual surfaces, and the height of the lingual cusps compared to the buccal cusps.

Mandibular First Molar

SIZE (Measured on the buccal surface)

Crown length (cervico-occlusal measurement on buccal surface of mesiobuccal cusp)	7.4 mm.
Crown width (greatest mesiodistal measurement)	11.0 mm.
Root length (measured from cervical line to apex of mesial root)	13.6 mm.

BUCCAL ASPECT (Examine several extracted teeth as you read. Hold crowns up, roots down.)

1. CROWN (Fig. 9-1a)

    Shape

    The crown is larger mesiodistally than cervico-occlusally.

There are 5 cusps:

The *mesiobuccal cusp*, the largest cusp on the buccal side.

The *distobuccal cusp*, slightly smaller than the mesiobuccal cusp.

The *distal cusp*, which is on the distobuccal angle of the crown, is the smallest of the cusps. (In Mongoloid peoples this cusp is often placed lingually. It may also be split into 2 parts by a fissure.[9])

The *mesiolingual cusp*, which is the highest of all of the cusps, is visible behind the mesiobuccal cusp.

The *distolingual cusp* is visible behind the distobuccal cusp.

The *mesial buccal groove* on the buccal surface separates the mesiobuccal from the distobuccal cusp.

Sometimes there is a deep *pit* at the cervical end of the mesial buccal groove. This is sometimes a site of caries. (This pit is seen frequently in Mongoloid teeth.[9])

The *distal buccal groove* separates the distobuccal cusp from the distal cusp. It is shorter than the mesial buccal groove.

The cervical line is often nearly straight across the buccal surface. Sometimes there is a point of enamel above the root bifurcation that dips down nearly into it. (This point of enamel is reported to occur in 90 per cent of Mongoloid peoples studied. Sometimes there is a dipping down of enamel on both the buccal and the lingual surfaces, and these extensions may meet in the root bifurcation.[7,9])

The mesial side of the crown is nearly straight, or slightly concave, from the cervical line to the contact area.

The distal side is nearly straight, or slightly convex, from the cervical line to the contact area.

The occlusal surface slopes cervically from mesial to distal.

Variations:

The mandibular first molar sometimes has an extra cusp on the buccal surface of the mesiobuccal cusp, about in the middle third of the crown. Studies have shown this to occur frequently in the Pima Indians of Arizona[2] and in Indian (Asian) populations.[3,5] Such an extra cusp has also been found on second and third molars (Figs. 9-4 and 9-5).

Contact areas

Mesial: About the junction of the occlusal and the middle thirds.

Distal: On the distal cusp, near the middle of the middle third.

2. ROOTS

There are a *mesial root* and a *distal root*. The mesial root is sometimes longer than the distal root.

The *root bifurcation* is near the cervical line; the root trunk is short. (In Mongoloid peoples there is usually a longer root trunk.[9])

There is a depression on the trunk above the bifurcation of the roots.

In the mandibular first molar there is often a ridge of cementum crossing the space in the root bifurcation in a mesiodistal direction.[4]

The roots on the first molar are usually widely separated.

The apical half of the roots curves distally from about the middle of each root.

The crest of curvature of the mesial side of the mesial root may extend mesial to the mesial surface of the crown.

The curvature of the roots is enough that:

The apex of the distal root often lies distal to the distal surface of the crown.

The apex of the mesial root often is mesial to the line of the mesial buccal groove.

The distal root is more pointed than the mesial root.

From the buccal aspect it is possible to see the distal side of the mesial root because of the way it is twisted on the trunk.

Variations:

Occasionally the mesial root is divided into a mesiobuccal and a mesiolingual root, making 3 roots on the mandibular first molar. It is reported that this condition is found in 10 to 20 per cent of the mandibular first permanent molars in Eskimos.[8] In Mongoloid peoples 10 per cent of the mandibular first molars have an additional distolingual root, and sometimes the mesial root is bifurcated, giving a 4-rooted first molar.[10] It is reported that in both deciduous and permanent dentitions 3-rooted mandibular molars occur frequently in Mongoloid peoples, but rarely in European groups.[9,11]

LINGUAL ASPECT

1. CROWN

The *mesiolingual* and the *distolingual* cusps are on the lingual side of the crown. The mesiolingual cusp is the longer.

They are both slightly longer than the buccal cusps.

Both are somewhat pointed at the tip.

The *lingual groove* separates the mesiolingual from the distolingual cusp.

The cervical line may be irregular.

The mesial side of the crown is nearly straight from the cervical line to the contact area.

2. ROOTS

The root trunk appears longer on the lingual side than on the buccal because the cervical line is more occlusal in position on the lingual than on the buccal surface.

From the lingual aspect it is possible to see the mesial surface of the mesial root due to the way it is twisted on the trunk.

MESIAL ASPECT

1. CROWN

The crown appears to be tilted lingually.

The buccal surface is very convex in the cervical third, then only slightly curved and inclined lingually in the middle and occlusal thirds.

The crest of curvature of the buccal surface is in the cervical third. It is called the *buccal cervical ridge.*

The lingual surface of the crown appears nearly straight in the cervical third, but is directed slightly lingually from the cervical line.

The crest of curvature of the lingual surface of the crown is usually in the middle third. Sometimes it is at the junction of the middle and occlusal thirds.

The *mesiolingual cusp* is longer than the distolingual cusp.

The tip of the mesiolingual cusp is often on a line with the lingual surface of the root.

The *mesial marginal ridge* is concave.

The cervical line is directed occlusally from buccal to lingual.

2. ROOTS

The mesial root is broad buccolingually and blunt at the apex.

There is often a *deep depression* on the mesial surface of the mesial root extending from the cervical line to the apex.

The mesial root has 2 root canals: 1 buccal and 1 lingual. Sometimes this root is divided into a buccal and lingual part.[10]

DISTAL ASPECT

1. CROWN (Fig. 9-1c)

The curvatures of the buccal and lingual surfaces are similar in appearance to those seen from the mesial aspect.

The *distal cusp* is at the distobuccal angle of the crown.

The *distobuccal cusp* is just mesial to the distal cusp. It is longer than the distal cusp.

The *distolingual cusp* is longer than the distobuccal cusp.

Because of the sloping of the occlusal surface and the distal tilt to the crown, part of the occlusal surface can be seen from the distal aspect.

The *distal marginal ridge* is V-shaped, often with a groove crossing it just distal to the distal cusp.

The crown is narrow on the distal side: some of the buccal and lingual surfaces can be seen.

The cervical line is nearly straight and slants occlusally from buccal to lingual. It is less occlusal in position on the distal surface than on the mesial surface.

2. Roots

The distal root is not quite so broad as the mesial root, and it is more pointed at the apex.

On some teeth the distal surface of the distal root is convex; on other teeth there may be a shallow longitudinal depression.

OCCLUSAL ASPECT (Figs. 9-1b and 9-2)

To follow this description the tooth should be held in such a position that the observer is looking exactly parallel to the axis line of the tooth. Because of the inclination lingually of the buccal surface, a considerable portion of the buccal surface is visible when the tooth is in this position.

There are 5 cusps:

Mesiobuccal, distobuccal, mesiolingual, distolingual, distal.

The crown is broader on the buccal than on the lingual side, and broader on the mesial than on the distal side.

The buccal surface is very convex; the lingual surface is slightly convex.

The mesial surface is very slightly convex, and slants distally from buccal to lingual.

The distal surface is convex, and slants mesially from buccal to lingual.

There are 3 fossae: the *central fossa*, the *mesial triangular fossa*, the *distal triangular fossa*.

There may be a deep *pit* at the bottom of one or all of these fossae.

There are several principal grooves:

The *central groove* extends from the central fossa mesially to the mesial triangular fossa and distally to the distal triangular fossa.

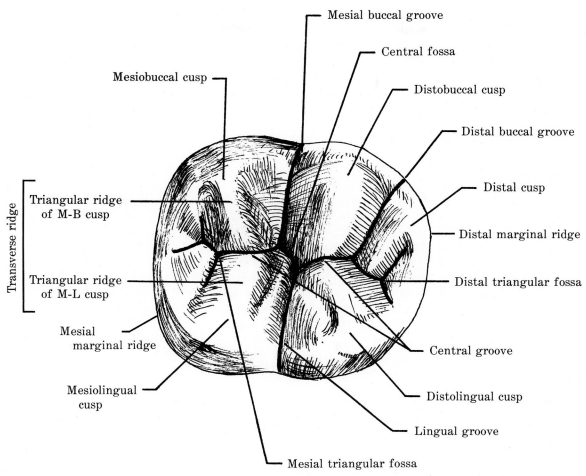

FIG. 9-2. Occlusal surface of mandibular right first molar. Drawn from a tooth model.

The *mesial buccal groove* starts in the central groove just mesial to the central fossa and extends between the mesiobuccal and the distobuccal cusps onto the buccal surface.

The *distal buccal groove* starts in the central groove between the central fossa and the distal triangular fossa and extends between the distobuccal and the distal cusps onto the buccal surface.

The *lingual groove* starts in the central fossa and extends lingually between the mesiolingual and the distolingual cusps onto the lingual surface.

There are numerous supplemental grooves.

The *triangular ridges* of the lingual cusps are longer than the triangular ridges of the buccal cusps.

The triangular ridges of the mesiobuccal cusp and mesiolingual cusp join to form a *transverse ridge*. And the triangular ridges of the distobuccal cusp and distolingual cusp join to form a *transverse ridge*.

The triangular ridge of the distal cusp does not join another triangular ridge.

The *central fossa* is about in the center of the occlusal surface.

Contact areas

Mesial: About in the center buccolingually.

Distal: On the distal cusp.

Variations:

The pattern of the grooves on the occlusal surface of the mandibular molars shows considerable variation. Studies have been made of the occlusal anatomy of these teeth, both in ancient and in modern man. Three principal types of occlusal groove pattern have been described: type Y, in which the central groove forms a Y figure with the lingual groove; type +, in which the central groove forms a + figure with the mesiobuccal and the lingual grooves; and type X, in which the occlusal grooves are somewhat in the form of an X.[6] If you will examine a collection of extracted mandibular molar teeth you will see some of these variations.

Mandibular Second Molar

SIZE (Measured on the buccal surface)

Crown length (cervico-occlusal measurement on the buccal surface of the mesiobuccal cusp)	6.9 mm.
Crown width (greatest mesiodistal measurement)	10.5 mm.
Root length (measured from the cervical line to the apex of the *mesial* root)	13.2 mm.

GENERAL CHARACTERISTICS

There are usually 4 cusps: *Mesiobuccal, distobuccal, mesiolingual, distolingual.*

The crown is usually smaller than the crown of the first molar in the same mouth. (It is reported that in the Bantu family of people in Africa, and sometimes in Eskimos, the mandibular molars often increase in size from first to third so that the third molar is the largest and the first molar is the smallest. This is not the most frequent order of size found in European peoples, but this sequence has been reported as occurring in a sample Ohio population. It is reported to occur also in Pima Indians. (Ref. 2, Chapter 8.)

The roots are less broad than on the first molar, and less widely separated.

BUCCAL ASPECT

1. CROWN (Fig. 9-3a)

   Shape

   As in the first molar, the crown is larger mesiodistally than cervico-occlusally.

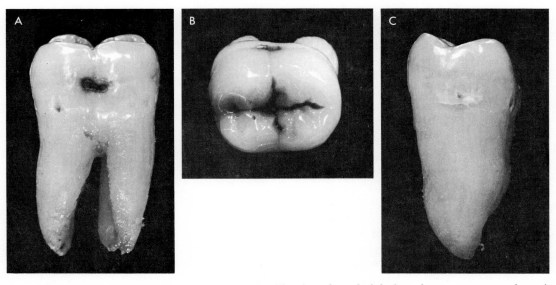

FIG. 9-3a. Mandibular right second molar. Buccal surface. The pit at the end of the buccal groove appears to be carious. **b.** Occlusal surface. Buccal side at top; mesial at left. This would be an interesting tooth to grind down in order to examine the deep grooves for caries. **c.** Distal surface. Buccal side at right.

All 4 cusps are visible from the buccal side. The tips of the *mesiolingual* and *distolingual cusps* are seen behind the mesiobuccal and distobuccal cusps.

The *mesiobuccal cusp* is slightly larger than the *distobuccal cusp.*

The *buccal groove* separates the mesiobuccal and the distobuccal cusps.

Sometimes there is a *pit* at the cervical end of the buccal groove.

The cervical line is nearly straight. It may have a point at the bifurcation of the root.

The mesial side of the crown is nearly straight from the cervical line to the contact area.

The distal side is convex from the cervical line to the contact area.

The crown tilts distally.

The occlusal surface slopes cervically from mesial to distal.

Contact areas

Mesial: About the junction of the middle and occlusal thirds.

Distal: The center of the middle third of the total crown length.

2. ROOTS

The root bifurcation extends to near the cervical line. The trunk is short. (In Mongolians the root trunk may be long, or the roots may be completely fused.[9])

The roots taper apically. Both roots are more pointed than the roots of the first molar.

Often the apex of both roots is directed toward the center line of the tooth.

Sometimes both roots curve distally.

Sometimes the distal root is slightly longer than the mesial.

LINGUAL ASPECT

The *mesiolingual* and *distolingual cusps* are visible.

The *lingual groove* separates these 2 cusps.

The cervical line slopes apically from mesial to distal.

The crown is slightly narrower on the lingual side than on the buccal.

The short root trunk has a depression above the bifurcation.

MESIAL ASPECT

1. CROWN

The appearance is very similar to that of the first molar.

The mesiolingual cusp is the longest of the cusps.

The mesiobuccal cusp is shorter than the mesiolingual.

The crest of curvature of the lingual surface of the crown is about in the middle of the middle third.

The crest of curvature of the buccal surface of the crown is close to the cervical line.

Occlusal to the buccal crest of curvature the crown slants lingually.

The tip of the mesiobuccal cusp is usually lingual to the line of the buccal side of the mesial root.

The tip of the mesiolingual cusp is usually about in line with the lingual outline of the root.

The *mesial marginal ridge* is concave.

2. ROOTS

The mesial root is broad and has a longitudinal depression on its mesial surface.

It is more pointed at the apex than in the first molar.

The mesial root often has 2 root canals.

DISTAL ASPECT

1. CROWN (Fig. 9-3c)

> The tips of the mesiobuccal and mesiolingual cusps can be seen behind the distobuccal and distolingual cusps.

> The distobuccal cusp is the shortest of the 4 cusps.

> Because the crown is narrower on the distal side than on the mesial side, the buccal and the lingual surfaces are visible.

> Because the crown tilts distally, much of the occlusal surface is visible.

> The *distal marginal ridge* is concave.

OCCLUSAL ASPECT (Fig. 9-3b)

> To follow this description the tooth must be held so that the observer is looking exactly parallel to the axis line of the tooth. Because of its inclination lingually, much of the buccal surface of the crown is visible when the tooth is in this position.

> The crown is slightly longer mesiodistally than buccolingually.

> The crown is narrower on the distal than on the mesial side, and narrower on the lingual than on the buccal side.

> The buccal surface of the mesiobuccal cusp often has a prominent bulge.

> The mesial side of the crown is nearly straight; the distal side is convex.

> The triangular ridges of the mesiobuccal and mesiolingual cusps meet to form a transverse ridge; and the triangular ridges of the distobuccal and distolingual cusps also form a transverse ridge.

> *Fossae:*

>> The *central fossa* is about in the center of the occlusal surface.

>> The *mesial triangular fossa* is just inside the mesial marginal ridge.

>> The *distal triangular fossa* is just inside the distal marginal ridge.

>> A *pit* may be present in any of the fossae.

> *Grooves:*

>> The *central groove* extends from the central fossa to the mesial triangular fossa and from the central fossa to the distal triangular fossa.

>> The *buccal groove* separates the mesiobuccal and distobuccal cusps.

>> The *lingual groove* separates the mesiolingual and the distolingual cusps.

>> There are sometimes supplemental grooves.

Contact areas:

Mesial: Near the junction of the buccal and middle thirds.

Distal: Near the junction of the buccal and middle thirds.

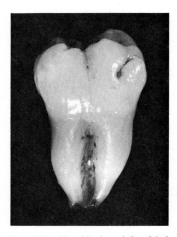

FIG. 9-4. Mandibular right third molar. Buccal surface. An extra cusp is on the buccal surface of the mesiobuccal cusp.

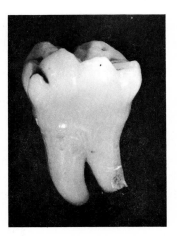

FIG. 9-5. Mandibular left third molar. Buccal surface. There is an extra cusp on the mesial surface of the mesiobuccal cusp.

Mandibular Third Molar

1. CROWN (Figs. 9-4 and 9-5)

The crown of the mandibular third molar sometimes resembles the crown of the mandibular second molar and sometimes the crown of the mandibular first molar. Again, it may bear little resemblance to either. This tooth is extremely variable.

The third molar usually has numerous supplemental grooves on the occlusal surface which produce a wrinkled appearance.

In Caucasians the third molar is usually the smallest of the mandibular molars.

2. ROOTS

There are 2 roots, mesial and distal, but often these are fused together. If the roots are separate, they are usually more pointed at the apex than the roots of the other molars. Often they curve distally more than in other molars.

References

1. Dahlberg, A. A., *The Dentition of the American Indian.* In Laughlin, W. S. (Ed.), *The Physical Anthropology of the American Indian,* New York, The Viking Fund, Inc., 1949.
2. Dahlberg, A. A., *The Evolutionary Significance of the Protostylid.* Amer. J. Phys. Anthrop., *8:*NS:15, 1950.
3. Dahlberg, A. A., *Geographic Distribution and Origin of Dentitions.* Int. Dent. J. *15:*348-355, 1965.
4. Everett, F. G., Jump, E. G., Holder, T. D., and Williams, G. C., *The Intermediate Bifurcation Ridge: A Study of the Morphology of the Bifurcation of the Lower First Molar.* J. Dent. Res., *37:*162, 1958.
5. Hellman, M., *Racial Characters of Human Dentition.* Proc. Amer. Philosoph. Soc., Vol. *67,* No. 2., 1928.
6. Jorgensen, K. D., *The Dryopithecus Pattern in Recent Danes and Dutchmen.* J. Dent. Res., *34:*195, 1955.
7. Masters, D. H., and Hoskins, S. W., *Projection of Cervical Enamel into Molar Furcations.* J. Periodont., *35:*49-53, 1964.
8. Scott, J. H., and Symons, N. B. B., *Introduction to Dental Anatomy,* London, E. and S. Livingstone, Ltd., 1958.
9. Tratman, E. K., *A Comparison of the Teeth of People. Indo-European Racial Stock with Mongoloid Racial Stock.* The Dental Record, *Vol. LXX.* No. 2., pp. 31-53, Feb. 1950, and No. 3, pp. 63-88, Mar. 1950.
10. Tratman, E. K., *Three-Rooted Lower Molars in Man and Their Racial Distribution.* Brit. Dent. J., *64:*264-274, 1938.
11. Turner, C. G., *Three-Rooted Mandibular First Permanent Molars and Question of American Indian Origin.* Amer. J. Phys. Anthrop. *34:*299-242, 1971.

Notes

Examine the mouth of a number of your associates and see how many still have the third molar teeth.

Of the third molars you examine, do you find any with a distal cusp resembling that of the first molar?

Of the individuals you examine, what proportion of those over age 18 have lost one or both mandibular first molars?

10

Tissues of Teeth and Pulp
Cavities of Permanent Teeth

I. PARTS OF A TOOTH

A. ANATOMIC DESCRIPTION (Figs. 2-3, 10-1)

1. *Anatomic crown*—the part of a tooth that has an enamel surface.
2. *Anatomic root*—the part of a tooth that has a cementum surface.
3. *Pulp cavity*—the space in the center of the crown and root which contains the tooth pulp.

B. CLINICAL DESCRIPTION (Fig. 1-1)

1. *Clinical crown*—the part of a tooth that is exposed in the oral cavity. This may be larger or smaller than the anatomic crown: it may include all of the anatomic crown and some of the anatomic root if there has been recession of the gingiva; or it may include only part of the anatomic crown if the cervical part of the crown is still covered by gingiva.

    You will find an example of a *small clinical crown* if you look at the maxillary central incisors of a 9- or 10-year-old child: the cervical part of the anatomic crown will be covered by gingiva.

    You probably will find an example of a *large clinical crown* if you look at the teeth of a 60-year-old (or younger) person: the cervical cementum of the anatomic root will probably be exposed.

2. *Clinical root*—the part of a tooth that is embedded in the jaw and is not exposed to the oral cavity. It may be greater or less than the anatomic root.

II. TISSUES OF A TOOTH

A. THE 4 TISSUES OF A TOOTH ARE:

1. *Enamel*
2. *Dentin* —hard (calcified) tissues, in order of decreasing hardness.
3. *Cementum*

4. *Pulp*—a soft (not calcified) tissue.

B. LOCATION OF THESE TISSUES (Fig. 10-1)

1. *Enamel*—makes up the surface of the anatomic crown.
2. *Cementum*—makes up the surface of the anatomic root.

97

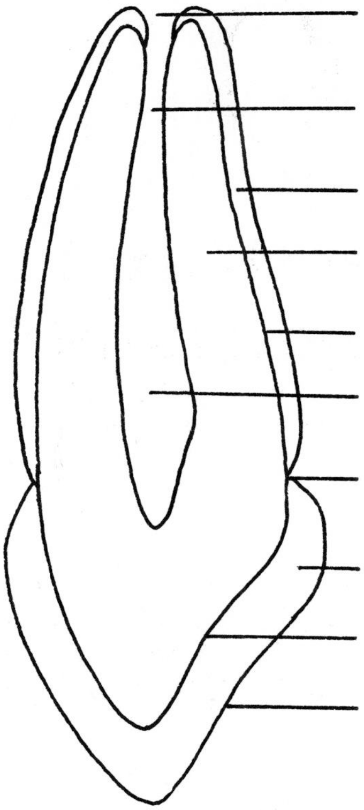

Apical foramen

Root canal

Cementum

Dentin

Cementodentinal
junction

Pulp chamber

Cementoenamel
junction

Enamel

Dentinoenamel junction

Lingual surface
of crown

FIG. 10-1.   Diagram of a maxillary anterior tooth sectioned longitudinally through the center to show the distribution of the tooth tissues and the shape of the pulp cavity.

3. *Dentin*—in the crown and the root, beneath the enamel and cementum.
4. *Pulp*—in the center part of the tooth, in the pulp cavity.

C. JUNCTIONS OF THESE TISSUES

1. *Dentinoenamel junction*
2. *Cementoenamel junction*
3. *Cementodentinal (dentinocemental) junction*

III. PULP CAVITIES OF THE TEETH

A. DIVISIONS OF PULP CAVITIES (for purpose of description)

1. *Pulp chamber*—in the crown of anterior teeth; partly in the crown, but mostly in the cervical part of the root of posterior teeth. Its wall is the inside surface of the dentin. There is one pulp chamber in each tooth.

2. *Root canal (pulp canal)*—in the root of a tooth. There is a canal in each root if there is more than one root. Some roots have two canals. The root canal is a continuation of the pulp chamber.

Openings to the outside:
Apical foramen—at or near the root apex (Figs. 10-7 and 10-14).

Accessory foramina—usually near the apex, but may be on one side of the root or even, in molars, at the root furcation.

B. Variation in Shape of Pulp Cavities

  1. *Of pulp chamber*

     Varies with: age of tooth, function, history (caries, attrition, tooth damage).

     In young teeth it resembles the shape of the crown surface: it has horns extending into the cusps, and usually it is constricted at the cervix.

  2. *Of root canal*

     Varies with age: gets smaller with age due to the addition of dentin on the internal wall.

Note:

   Using an abrasive stone on a dental lathe in the manner described in the *Introduction*, remove different sides of each kind of tooth and carefully examine the pulp cavities. Sketching the cavity outlines as you see them from different planes of cutting is helpful.

   In incisor and canine teeth either the mesial or distal side should be removed from some teeth and the facial or lingual side removed from others so that the shape of the pulp cavity can be seen from both the mesiodistal and labiolingual planes.

   In a premolar tooth the removal of either the mesial or distal side will expose the pulp horns which extend into the cusps; and in another specimen removal of the buccal and lingual sides to the level of the buccal and lingual cusp tips will reveal the shape of the pulp cavity in a mesiodistal plane.

   In molar teeth some specimens should be prepared by removal of the buccal surface and some by removal of the lingual surface. Some teeth should have the mesial surface removed to the level of the tips of the mesiobuccal and mesiolingual cusps, and others should have the distal surface removed to the level of the distobuccal and distolingual cusps. Some molar teeth should have the occlusal surface removed so that the openings to the root canals on the floor of the pulp chamber can be seen.

IV. SHAPE OF PULP CAVITIES IN SOUND, YOUNG TEETH

  A. In Incisors and Canines (as seen when a tooth is ground to expose pulp cavity) (Figs. 10-2, 10-3, 10-4, 10-5, 10-6, 10-7)

  1. *Pulp chamber*

     a. Cut labiolingually: the pulp chamber tapers to a point toward the incisal edge (Fig. 10-5).

     b. Cut mesiodistally:
        In maxillary and mandibular central and lateral incisors, the pulp chamber is broad and may have a suggestion of mesial and distal horns (Fig. 10-2).

        In maxillary and mandibular incisors, a young tooth may show the configuration of mamelons at the incisal border of the pulp wall.

        In maxillary and mandibular canines, the incisal wall of the pulp chamber is usually rounded.

2. *Root canal*

    a. In newly emerged teeth:
        The root is not completely formed.
        The root canal is very large.
        The apical opening may be the broadest part of the canal. (Try removing, by grinding, 1 side of a root of a newly erupted extracted third molar.)

    b. In fully formed young teeth:
        The root is completely formed.
        The root canal tapers toward the apical foramen.
        The apical foramen is relatively small (Fig. 10-2).

    c. The number of root canals in each anterior tooth:
        Maxillary and mandibular incisors—1 canal.
        Maxillary canines—1 canal. (Very rarely there are 2 canals, facial and lingual.)
        Mandibular canines—1 canal. (Occasionally there are 2 roots and 2 canals, facial and lingual.)

B. IN PREMOLARS (as seen when the tooth is ground to expose the pulp cavity) (Figs. 10-8, 10-12, 10-13)

1. *Pulp chamber*

    a. Cut mesiodistally: the occlusal border is curved in each cusp similar to the curvature of the tooth surface.

    b. Cut buccolingually: the pulp horns are visible in each cusp.

    The buccal horn is longer than the lingual horn (Fig. 10-8).

    The pulp chamber often has the general outline of the tooth surface.

2. *Root canal*

    a. Maxillary first premolar: usually there are 2 roots, buccal and lingual.

    There are always 2 root canals, buccal and lingual (even if there is only 1 root).

    b. Maxillary second premolar: there are 1 root and 1 root canal.

    Sometimes there is a division into buccal and lingual canals, then a union at the apex, and 1 apical foramen.

    c. Mandibular first and second premolars ordinarily have 1 root and 1 root canal.

    Occasionally the mandibular first premolar has a buccal and a lingual root (Fig. 10-13).

    A mandibular second premolar with 2 roots is rare.

    d. The size of canals decreases with age due to deposition of additional dentin.

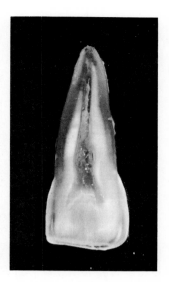

Fig. 10-2.   Maxillary central incisor; facial side removed. The high pulp horns and the broad root canal indicate that this is a young tooth.

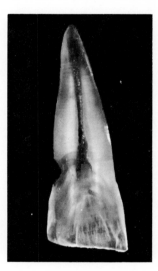

Fig. 10-3.   Maxillary central incisor; facial side removed. The pulp chamber is partly filled with secondary dentin and the root canal is narrower than in the tooth shown in figure 10-2. Also, the incisal edge is worn to a straight line. This is an older tooth than that in figure 10-2. The damage to the cervical part of the root on the distal side of the tooth has been there for some time because the underlying dentin has become altered. (See an oral histology textbook.)

Fig. 10-4.   Maxillary central incisor; mesial side removed. There is attrition on the incisal edge and secondary dentin in the incisal part of the pulp chamber. The root canal is moderately wide.

FIG. 10-5. Maxillary canine; mesial side removed. There is no extensive attrition on the incisal edge and the pulp cavity is still large.

FIG. 10-6. Mandibular lateral incisor; mesial side removed. Curvature of the root prevented cutting the pulp cavity in one plane so that the apical portion was lost. Notice how the pulp cavity extends in a narrow point toward the incisal edge. Extensive attrition on the incisal edge would not expose the pulp; secondary dentin would form in the incisal part of the pulp chamber and the pulp would be protected.

FIG. 10-7. Mandibular canine; mesial side removed. The pulp cavity is large; only at the incisal tip is there a little evidence of secondary dentin formation.

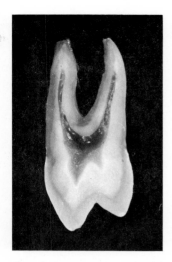

FIG. 10-8. Maxillary first premolar; mesial side removed. The curvature of the tips of the roots prevented cutting the root canals in 1 plane. The pulp horns are sharp; there is little secondary dentin. The buccal pulp horn is considerably longer than the lingual horn.

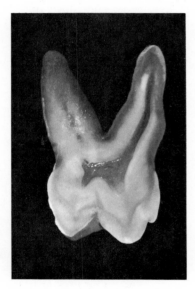

FIG. 10-9. Maxillary first molar; mesial side removed. The cut is through the center of the lingual root canal, but not through the center of the mesiobuccal canal. The pulp chamber is seen to open into the lingual root canal. There is an opening, a small hole, in the floor of the pulp chamber for each of the 3 root canals. Grind off the occlusal portion of the crown of a molar tooth and examine the floor of the pulp chamber.

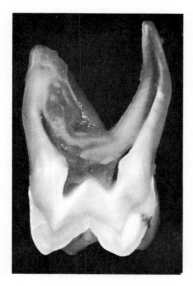

FIG. 10-10. Maxillary first molar; mesial side removed. Cut through the mesiobuccal and lingual root canals. Notice how the lingual canal takes the curvature of the root. The pulp chamber is mostly in the root trunk: only the mesiobuccal and mesiolingual pulp horns extend a little into the part of the tooth we define as the anatomic crown. You can determine the extent of the anatomic crown by the location of the cervical border of the enamel. In the groove where the small cusp of Carabelli is attached to the mesiolingual cusp there is an area of dental caries.

C. IN MAXILLARY MOLARS (*First and Second*) (as seen when teeth are cut in different planes) (Figs. 10-9, 10-10, 10-11)

1. *Pulp chamber*

There is a pulp horn in each cusp.

The pulp chamber is broader buccolingually than mesiodistally.

Often it is constricted near the floor of the chamber.

The floor* of the pulp chamber is apical to the cervical line: it is located in the root trunk.

The floor is level in young teeth. It may be convex in older teeth due to deposition of additional dentin with age.

The root canals open from the floor of the pulp chamber.

2. *Root canals*

In mesiobuccal and distobuccal roots, the canals are narrow.

In the lingual root the canal is larger and more easily accessible from the floor of the pulp chamber than in the other 2 roots.

There may be several accessory foramina. There may be a foramen in the furcation of the roots that opens from the floor of the pulp chamber.

D. IN MANDIBULAR MOLARS (*First and Second*) (as seen when teeth are cut in different planes) (Figs. 10-14, 10-15)

1. *Pulp chamber*

The pulp chamber has a pulp horn in each cusp.

The mesial horns are longer than the distal horns (Fig. 10-15).

The chamber is broader mesiodistally than buccolingually. (This is the reverse of the maxillary molars.)

Often there is a constriction near the floor of the chamber.

The root canals open from the floor of the pulp chamber.

2. *Root canals*

a. In the mandibular first molar

The mesial root usually has 2 canals: 1 buccal, 1 lingual. Sometimes these join and have a common foramen at the apex.

The distal root usually has 1 canal.

---

* In both maxillary and mandibular molar teeth the occlusal wall of the pulp chamber is called the *roof*, and the wall next to the root is called the *floor*.

b. In the mandibular second molar

The mesial root usually has only 1 canal, broad buccolingually and narrow mesiodistally.

The distal root has 1 canal.

c. All canals are narrow when the root development is completed.

V. SHAPE OF PULP CAVITIES IN OLDER TEETH

A. DENTIN FORMATION

1. *Dentin* is a product of tooth pulp.

2. *Dentin formation* may continue as long as the pulp is intact—dentin forms on the wall of the pulp cavity.

3. *Dentin formation* may be stimulated by aging, attrition, and caries (Fig. 10-11).

B. RESULTS OF CONTINUED DENTIN FORMATION

1. *Reduction in size of the pulp chamber:* it may become entirely filled in some cases.

2. *Narrowing of pulp canal* (Fig. 10-3).

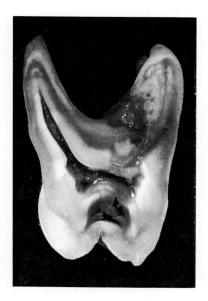

FIG. 10-11. Maxillary first molar; mesial side removed. If you grind a number of teeth to expose the pulp cavity you will easily see why some of them were lost. In this tooth the caries lesion has spread under the enamel from its point of entry (not seen here) to undermine a large part of the enamel, penetrate through a broad area of dentin, and reach the pulp. Examination of caries lesions by the simple process of judiciously grinding off one side of the tooth makes an extremely interesting study.

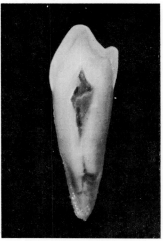

FIG. 10-12. Mandibular first premolar; distal side removed. Root curvature prevented cutting the root canal in 1 plane. The pulp horn in the buccal cusp is large; in the lingual cusp it is small.

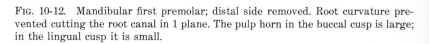

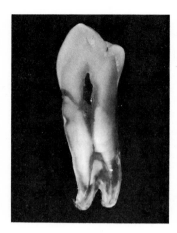

Fig. 10-13. Mandibular first premolar with root and root canal divided near the apex. Mesial side removed.

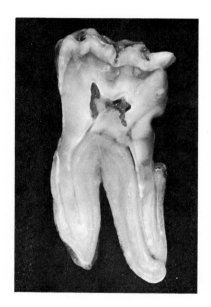

Fig. 10-14. Mandibular first molar; buccal side removed. The apical foramen of the distal root is on the distal side of the root, not at the root tip. Sometimes there are several accessory foramens on a root. The roof of the pulp chamber is about at the same level as the cervical border of the enamel on the mesial side of the tooth; only the pulp horn extends into the anatomic crown; most of the pulp chamber is located in the root trunk.

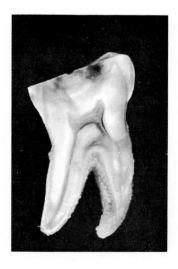

Fig. 10-15. Mandibular first molar. Lingual side removed. There appears to be caries in the enamel above the mesiolingual pulp horn; but it has penetrated the dentin only slightly. Again notice that the roof of the pulp chamber is about at the level of the cervical line. The pulp horns extend occlusal to the cervical line; the rest of the pulp chamber is in the root trunk.

Notes

It is interesting to keep a record of peculiarities you see in teeth you have ground to expose the pulp cavities.

Watch carefully for *accessory foramina* (foramina in places other than the root tip). Look particularly on the sides of the root in the apical third and in the furcation of multirooted teeth.

11

The Primary Dentition

Your best specimens for the study of crown morphology can probably be found in the mouth of a cooperative little friend (2 to 6 years) willing to open his mouth wide, long, and often to permit your examination. Extracted primary teeth, even those with resorbed roots, are sometimes obtained with difficulty. Tooth models, if available, are helpful.

PRIMARY TEETH emerge between the ages of 6 months and 2 years (Fig. 11-1). Beginning at the age of 6 or 7 years these teeth are gradually replaced by the teeth of the permanent dentition.

Sometimes primary teeth are called *deciduous teeth*. The word *deciduous* comes from a Latin word meaning *to fall off*. Deciduous teeth fall off, or are shed.

DENTAL FORMULA FOR THE HUMAN PRIMARY DENTITION

$I\frac{2}{2}\,C\frac{1}{1}\,M\frac{2}{2}$ = 10 teeth on either side. A total of 20 teeth.

When primary teeth are replaced by teeth of the permanent dentition the primary molars are replaced by the permanent premolars. The permanent molars have no predecessors in the primary dentition. Each mouth quadrant may be represented in this way:

Primary teeth	Incisors $\frac{2}{2}$	Canines $\frac{1}{1}$	Molars $\frac{2}{2}$	
Permanent replacements	Incisors $\frac{2}{2}$	Canines $\frac{1}{1}$	Premolars $\frac{2}{2}$	Molars $\frac{3}{3}$

IMPORTANT FUNCTIONS OF SOUND PRIMARY TEETH:

1. Efficient mastication of food. (Missing teeth may be one reason children reject foods that are difficult to chew.)

2. Avoidance of infection.

3. Maintenance of space for the emergence of permanent teeth.

ERUPTION TIME OF PRIMARY DENTITION

The first primary teeth to erupt are usually the mandibular central incisors, at about 6 months.

IDENTIFICATION OF TEETH BY NUMBER AND TIME OF EMERGENCE INTO ORAL CAVITY

PRIMARY DENTITION

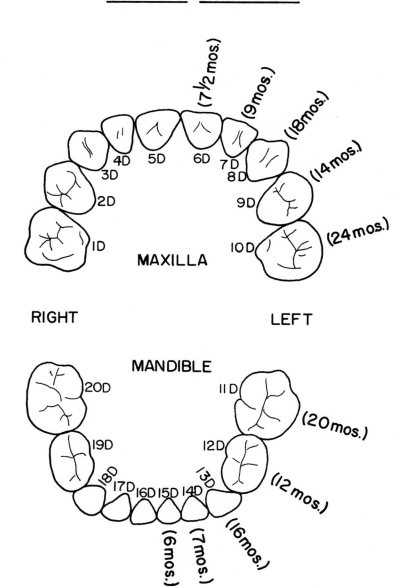

MAXILLA

RIGHT   LEFT

MANDIBLE

EMERGENCE TIME IN MONTHS IS INDICATED BY FIGURES IN PARENTHESES.

FIG. 11-1. A diagram of the maxillary and mandibular primary dental arches. The numbers in parentheses indicate the ages at which each tooth may be expected to emerge into the oral cavity. The numbers 1D through 20D inside the arches indicate one of several ways in which the teeth may be coded (rather than named) for purposes of record-keeping. Starting at 1D, the teeth are right second molar, first molar, canine, lateral incisor, and central incisor (which is 5D). The remaining teeth are similarly coded.

The last primary teeth to erupt are the maxillary second molars, at about 24 months.

The first permanent teeth to erupt are usually the maxillary or mandibular first molars. They appear distal to the second primary molars at about 6 years. It is important to recognize that these are permanent teeth. They are commonly called 6-year molars.

Permanent incisors begin to replace primary incisors at about 6 or 7 years.

The last primary teeth to be lost are the mandibular or maxillary second molars, at about 11 to 12 years. These teeth are replaced by the second premolars of the permanent dentition.

The complete primary dentition is in the mouth from 2 years to 6 or 7 years of age, and no permanent teeth are present.

A *mixed dentition* (both primary and permanent teeth) is present from 6 or 7 years to 12 years.

TIME OF EMERGENCE OF PRIMARY TEETH*

	Maxillary	*Mandibular*
Central Incisor	7½ mos.	6 mos.
Lateral Incisor	9 mos.	7 mos.
Canine	18 mos.	16 mos.
First Molar	14 mos.	12 mos.
Second Molar	24 mos.	20 mos.

CHARACTERISTICS OF PRIMARY TEETH

Primary teeth are smaller in size than permanent teeth and often white in color.

The anterior teeth:

The crown is narrow at the cervix.
Usually there are no depressions on the labial surface of the crowns of the incisors.
The cervical ridge on the facial surface is prominent.
The root is long in proportion to the crown length, and is narrow.

The posterior teeth:

There are few grooves or depressions in the crowns.
The cervical ridge is very prominent.
The root furcation is near the crown and there is little root trunk.
The roots are widely spread.

*Dates given in Orban, B. J. (Ed.), *Oral Histology and Embryology*, 5th ed., edited by H. Sicher, St. Louis, Mosby, 1962.

The roots:

> The root of a primary tooth is not completely formed for a year or more after the emergence of the crown into the mouth. And the intact root is short-lived: after a year or two it starts to resorb, usually at the apex, or on one side near the apex, as the crown of the permanent tooth which is to replace it becomes large enough to infringe upon the primary root. Increasing loss of root attachment due to root resorption results in the loosening of the primary tooth, and it *falls off* of the jaw. Usually the crown of the permanent successor is close to the surface, ready to emerge.

Primary Maxillary Central Incisors

LABIAL ASPECT

The crown is wider than it is long, but it is narrow near the cervix.

The labial surface is smooth; usually there are no depressions.

The root is longer relative to the crown length than in permanent teeth: it is about twice the length of the crown. On extracted or shed teeth which are obtainable for study there is usually some root resorption. Often the entire root is gone.

The mesial contact area is near the mesioincisal angle; the distal contact area is in the incisal third (Fig. 11-2).

LINGUAL ASPECT

The cingulum is often proportionally large, so that the lingual fossa is in only the incisal third of the lingual surface.

The marginal ridges are often distinct.

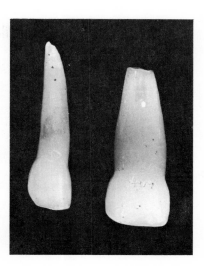

FIG. 11-2. Primary maxillary right lateral incisor (at left) and right central incisor. Labial surfaces. The central incisor crown is short and broad. There has been some resorption of the root tips on both teeth.

MESIAL AND DISTAL ASPECTS

The crown is wide labiolingually near the cervix due to the large cingulum.

The cervical curvature is greater on the mesial than on the distal surface.

INCISAL ASPECT

The labial surface is smooth.

The lingual surface becomes narrow at the cingulum.

Primary Maxillary Lateral Incisors

The lateral incisors differ from the central incisors in some ways:

The crown is longer than it is wide. (This is the reverse of the central incisor.)

The tooth is smaller than the central incisor (of the same dentition).

The distoincisal angle is more rounded (Fig. 11-2).

Primary Maxillary Canine

LABIAL ASPECT

When you examine a primary maxillary canine you will find that:

The mesial and distal surfaces are very convex.

The crown is constricted at the cervix.

Instead of a horizontal incisal edge the canine has a cusp which is often sharp.

The mesial and distal contact areas are about in the center of the crown cervico-occlusally.

The contour of the distal side of the crown is more rounded than that of the mesial side (Fig. 11-3).

LINGUAL ASPECT

Here you see the cingulum, the mesial and distal marginal ridges, and the cusp slopes.

On either side of the lingual ridge you will find a mesial and distal lingual fossa.

MESIAL AND DISTAL ASPECTS

The cervical third of the crown is much broader than in the incisors.

INCISAL ASPECT

> The contour is somewhat angular. It narrows toward the center of the labial surface and toward the center of the lingual surface.

> The smallness of these teeth compared to their permanent counterparts is surprising.

FIG. 11-3.   Primary maxillary left canine. Labial surface.

Primary Mandibular
Central and Lateral Incisors

The crowns resemble the permanent mandibular incisors, but they are very much smaller.

The thin roots are about twice the length of the crowns.

The lateral incisors are a little larger than the central incisors of the same dentition.

The labial surface is smooth; the lingual surface has a cingulum, two faint marginal ridges, and a slight lingual fossa (Fig. 11-4).

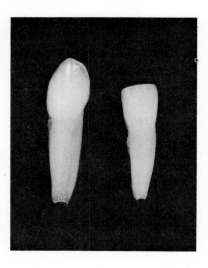

FIG. 11-4.   Primary mandibular right canine (at left) and right lateral incisor. Labial surfaces. The root tips have been resorbed. Some primary mandibular canines bear a closer resemblance than this one to their permanent successor.

Primary Mandibular Canine

The mandibular canine is slightly smaller than the maxillary canine in crown length and width, and is much smaller labiolingually.

It resembles the permanent canine in the shape of the crown. As in the permanent canine, the mesial cusp slope is shorter than the distal cusp slope (Fig. 11-4).

Primary Maxillary First Molars

This tooth is so different from a permanent maxillary first molar that at first inspection it is a surprise (Figs. 11-5a, b, c).

BUCCAL ASPECT

The border of the occlusal surface is scalloped, but the mesiobuccal and mesiolingual cusps are indistinct.

There is no buccal groove on the buccal surface.

The crown is narrow at the cervix.

The 3 roots are thin and widely spread, and there is very little root trunk: the furcation is close to the cervical line.

There is a prominent cervical ridge, particularly on the mesial side of the buccal surface. (Feel it.)

LINGUAL ASPECT

The mesiolingual cusp is the largest cusp on the tooth.

The distolingual cusp is inconspicuous, and may even be absent.

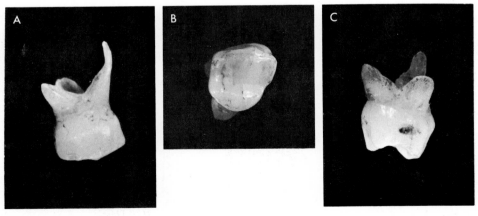

FIG. 11-5a.   Primary maxillary right first molar. Buccal surface. The mesiobuccal root is less resorbed than the distobuccal and lingual roots. The lingual root is barely discernible. **b.** Occlusal surface. Buccal side at top; mesial at right. The prominent cervical ridge on the mesiobuccal cusp gives the tooth an angular appearance. **c.** Distal surface. The roots are partly resorbed. Within the space enclosed by these 3 wide-spread roots the crown of the permanent maxillary first premolar develops.

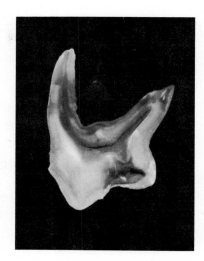

FIG. 11-6.   Primary maxillary first molar; mesial side removed. The root canals of the mesiobuccal root and of the lingual root (right side of picture) are exposed. An extensive caries lesion spread beneath the enamel of the lingual cusp (initial point of entry not seen here) has reached the pulp horn.

MESIAL ASPECT

The cervical third of the crown is much wider buccolingually than the occlusal third. The crown narrows toward the occlusal surface.

The cervical ridge on the buccal surface is very prominent.

The mesiolingual cusp is usually longer than the mesiobuccal cusp, and the cusp tip is sharper.

DISTAL ASPECT

The crown is narrower buccolingually on the distal side than on the mesial side.

The cervical ridge on the buccal surface is less prominent from the distal aspect than from the mesial.

OCCLUSAL ASPECT

The crown is broader on the buccal side than on the lingual, and broader on the mesial than on the distal.

But the occlusal surface is not the same shape as the crown contour: the occlusal surface would be square if the mesiolingual corner were not rounded.

Cusps: mesiobuccal, distobuccal, mesiolingual, distolingual (distolingual is sometimes absent).

Fossae and grooves:

Central fossa
Mesial triangular fossa
Distal triangular fossa
Central groove: connects central fossa and mesial triangular fossa.
Buccal groove (not extended to buccal surface) divides buccal cusps.

Primary Maxillary Second Molar

The primary second molar is considerably larger than the primary first molar (Figs. 11-7a, b, c, d).

It greatly resembles the permanent maxillary first molar.

It is similar in that:

From the occlusal aspect the crown appears square.

The cusps correspond to those of the permanent first molar: mesiobuccal, distobuccal, mesiolingual, distolingual, cusp of Carabelli.

It differs in that:

It is smaller.

The crown is narrow at the cervix and has a very prominent cervical ridge on the buccal surface.

The roots are widely spread. The root furcation is near the cervical line: there is practically no root trunk.

Primary Mandibular First Molar

This tooth is not easily compared to the teeth of the permanent dentition (Figs. 11-8a, b, c, d).

BUCCAL ASPECT

The mesial contour of the crown is nearly straight, while the distal side is very convex.

The crown is narrow near the cervix.

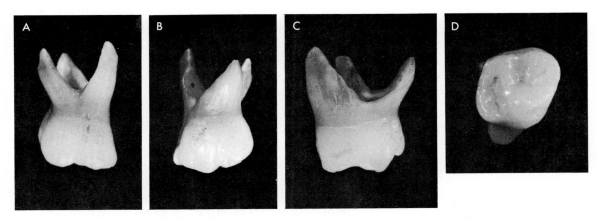

FIG. 11-7a. Primary maxillary right second molar. Buccal surface. **b.** Lingual surface. **c.** Mesial surface. Again notice the spread of the roots. The crown of the permanent maxillary second premolar develops in the space bounded by these roots. **d.** Occlusal surface. From this aspect the primary maxillary second molar resembles a miniature permanent first molar.

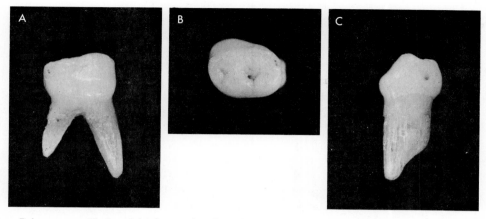

FIG. 11-8a. Primary mandibular right first molar. Buccal surface. The distal root has been considerably shortened by resorption. **b.** Occlusal surface. Buccal side at top; mesial at left. The large cervical ridge on the mesiobuccal cusp is conspicuous. **c.** Mesial surface. Again the cervical ridge is outstanding. Notice how it makes the crown appear to tilt lingually. If the root apex were not resorbed the root would be seen to taper to a blunt end.

The crown is shorter cervico-occlusally on the distal side than on the mesial.

There is no buccal groove between the mesiobuccal and distobuccal cusps.

The roots are widely spread and slender, and the furcation is close to the cervical line.

LINGUAL ASPECT

The mesial surface is visible when you look directly at the lingual side of the tooth, but the distal surface is not visible.

The mesiolingual cusp is larger than the distolingual, and is rounded.

There is a barely noticeable groove between the two lingual cusps.

MESIAL ASPECT

The cervical ridge on the buccal surface is prominent: the crown appears to lean lingually. All mandibular posterior teeth—primary and permanent—appear to lean lingually.

The buccal crown contour is nearly (not quite) flat from the buccal crest to the occlusal surface.

The mesiolingual cusp is longer than the mesiobuccal.

The mesial root is flat and square, sometimes with a depression its full length.

DISTAL ASPECT

The buccal cervical ridge is less prominent from the distal side than from the mesial.

The distobuccal and distolingual cusps are nearly the same height.

The distal marginal ridge is less prominent than the mesial.

The distal root is rounded and both thinner and shorter than the mesial root.

OCCLUSAL ASPECT

The mesiobuccal angle is prominent because of the cervical ridge on the buccal surface.

The mesiobuccal angle is an acute angle; the distobuccal angle is obtuse.

The cusps are often difficult to distinguish, but careful examination of an unworn tooth will reveal (in order of diminishing size) mesiolingual, mesiobuccal, distobuccal, distolingual cusps.

Grooves and fossae:

There are a mesial triangular fossa and a distal triangular fossa. A central groove separates the mesiobuccal and mesiolingual cusps.

There are a buccal groove and a lingual groove on the occlusal surface. These grooves may be indistinct, and they do not extend onto the buccal and lingual surfaces.

Primary Mandibular Second Molar

The primary mandibular second molar resembles the permanent mandibular first molar in many ways. It is smaller, however, and the roots are more slender and more widely spread. The root furcation is near the cervical line: there is very little root trunk (Figs. 11-9a, b, c).

BUCCAL ASPECT

Here there are 3 cusps of nearly equal size: mesiobuccal, distobuccal, distal.

The grooves are: mesial buccal and distal buccal.

The roots are about twice as long as the crown, and are widely spread.

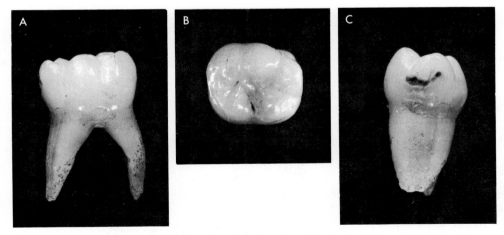

Fig. 11-9a. Primary mandibular right second molar. Buccal surface. The short root trunk and the wide-spread roots, as well as the small size, distinguish this tooth from the permanent mandibular first molar. The mesiobuccal, distobuccal, and distal cusps are often about the same size. **b.** Occlusal surface. Buccal side at top; mesial at left. **c.** Distal surface. Buccal side at right.

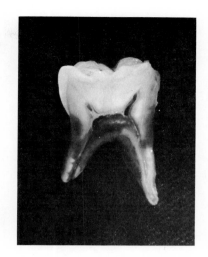

Fig. 11-10. Primary mandibular right second molar. Buccal side ground off to expose pulp cavity. An interesting feature is the long narrow shape of the pulp horns.

LINGUAL ASPECT

The mesiolingual and distolingual cusps are about the same size.

There is a lingual groove between them.

The lingual side of the crown is slightly narrower than the buccal.

MESIAL ASPECT

From this side this tooth resembles a permanent first molar.

One noticeable difference is that there is a more prominent cervical ridge on the buccal surface, causing a greater lingual tilt to the crown than in the permanent molar.

The mesial root is broad, flat, and blunt.

DISTAL ASPECT

The crown is narrower on the distal side than on the mesial.

The distal marginal ridge is lower (more cervical) than the mesial.

The distal root is broad and flat, but less blunt at the apex than the mesial. The early resorption of these roots makes it difficult to obtain specimens with complete roots. If the tooth has been shed naturally (not removed for some reason by a dentist) the roots are all but completely resorbed.

OCCLUSAL ASPECT

There are 5 cusps: mesiobuccal, distobuccal, distal, mesiolingual, distolingual, and each has a clear triangular ridge.

There are a central fossa and a mesial and distal triangular fossa.

There are a central groove, a mesial buccal groove, a distal buccal groove, and a lingual groove.

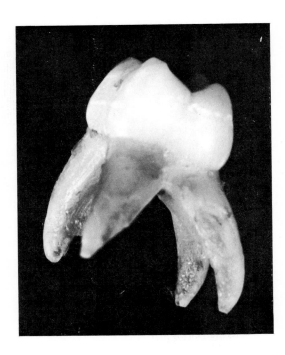

FIG. 11-11. Primary mandibular left second molar. This is the mesial side of the tooth; the buccal surface is at the right. The 4 roots are spread wide and little resorbed. No history is known about this specimen. Only when you hold this tooth in your hand and turn it around can you identify it as a mandibular molar.

There are also numerous small grooves.

Comparing the primary mandibular second molar and permanent mandibular first molar:

The primary second molar is smaller than the permanent first molar.

The 3 cusps on the buccal side are of nearly equal size. In the permanent first molar the distal cusp is smallest, the mesiobuccal is largest.

There is a prominent buccal cervical ridge on the primary molar.

The roots of the primary second molar are thinner and more widely spread, and the furcation is nearer the cervical line.

Pulp Cavities of Primary Teeth

Primary anterior teeth have pulp cavities that are similar in shape to the pulp cavities of the permanent teeth.

Primary molar teeth, when compared with permanent molars, have pulp chambers much narrower in vertical measurement relative to the size of the tooth. In permanent molars much of the pulp chamber is in the root trunk; in primary molars there is little, or almost no root trunk. In these teeth the pulp chambers are mostly, or entirely, in the tooth crown.

The pulp chambers of primary molar teeth usually have long, and often very narrow, pulp horns extending into the cusps.

Compare the relative size and shape of the pulp chambers of primary molars (Figs. 11-6 and 11-10) with the size and shape of those permanent molars (Figs. 10-10 and 10-15).

Sequence of Tooth Eruption

Of course, many variations occur, but the usual order of appearance of the teeth in the oral cavity is:

Primary Teeth	*Permanent Teeth*
Mandibular central	Maxillary and mandibular first molars
Mandibular lateral	Mandibular central
Maxillary central	Maxillary central and mandibular lateral
Maxillary lateral	Maxillary lateral
Mandibular first molar	Mandibular canine
Maxillary first molar	Maxillary first premolar
Mandibular canine	Maxillary second premolar and mandibular first premolar
Maxillary canine	
Mandibular second molar	Mandibular second premolar and maxillary canine
Maxillary second molar	Mandibular second molar
	Maxillary second molar
	Third molars

Notes

If you are fortunate enough to have a collection of primary teeth, study the morphology for variations.

Prepare a tooth of each type to study the shape of the pulp cavities. In the primary molar teeth, which as you see have little or no root trunk, how is the location of the pulp chamber different from the location of the pulp chamber in permanent molar teeth? Look carefully at both. (See figures 11-6 and 11-10.)

What parts of the roots of primary teeth seem to undergo resorption first?

Figure 11-11 shows the mesial side of a mandibular left second molar. There are 2 mesial roots and 2 distal roots. In the extracted primary teeth that you have had an opportunity to examine, have you found any other anomalies?

12

Oral Anatomy Related to Oral Function

ALIGNMENT OF THE TEETH

Alignment means the arrangement of the teeth in the dental arches.

CONTACT AREAS

Contact areas are the places on the approximal surfaces of tooth crowns where a tooth touches the tooth adjacent to it in the same arch when the teeth are in proper alignment. (Also called *contact points*.) (Figs. 12-1 and 12-2).

On different teeth contact areas may characteristically be in the incisal third, the middle third, or at the junction of the incisal or middle third. On no tooth is a contact area characteristically located in a position more cervical than the middle of the middle third of the crown.

The contact of each tooth with the adjacent teeth has important functions:

It stabilizes the tooth.
It helps prevent food impaction.
It protects the interdental papillae of the gingiva.

The *contact area* sometimes has to be distinguished from the *area of contact*, which may not be the same location if the teeth are not in good alignment, e.g., if a tooth is rotated or badly out of position.

INTERPROXIMAL SPACE

This is the triangular space between adjacent teeth cervical to their contact: the sides of the triangle are the approximal surfaces of the adjacent teeth, and the apex of the triangle is the area of contact of the 2 teeth. This space is occupied in young individuals by the *interdental papilla*. (See Figs. 1-1 and 13-3.)

EMBRASURE

An *embrasure* is the space between adjacent teeth where their approximal surfaces diverge from the area of contact facially, lingually, and occlusally (incisally) (Figs. 12-1 and 12-2).

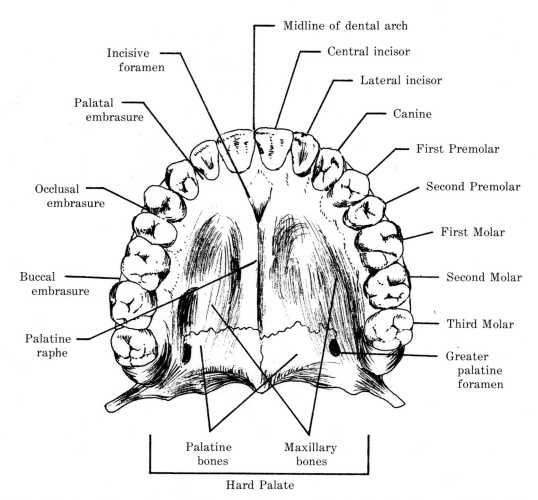

FIG. 12-1. A drawing of the maxillary dental arch and the bones of the hard palate. Compare this illustration with figure 12-2.

The lingual embrasures are ordinarily larger than the facial embrasures, since most teeth are narrower on the lingual side than on the facial side. Also, in posterior teeth, the contact areas are usually facial to the middle of the middle third. These locations are seen when the dental arch is examined from the occlusal view (Fig. 12-2.)

Embrasures act as spillways which direct food away from the gingiva.

GINGIVA

Gingiva is defined as that part of the oral mucuous membrane that is firmly attached to the alveolar process and to the cervical parts of the teeth and which surrounds the cervices of the teeth (Figs. 1-1 and 12-2).

INTERDENTAL PAPILLA

The *interdental papilla* is the part of the gingiva that is located in the interproximal space (Figs. 1-1 and 12-2).

In young individuals the interdental papilla fills the interproximal space and prevents food from becoming lodged between the teeth.

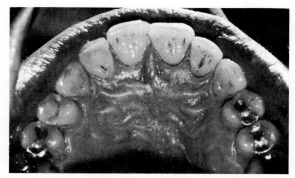

FIG. 12-2. Photograph of the anterior part of the palate. The palatal rugae are conspicuous. Interdental papillae fit into the palatal embrasure spaces. Notice the contact of each tooth with the ones adjacent to it. The morphology of the occlusal surface of the right first premolar is beautiful, and the linear depressions on the lingual surfaces of the incisors are distinct. The location of the restorations leaves little doubt about which areas are apt to be most susceptible to caries.

In older individuals the interdental papilla usually does not fill the interproximal space because of the recession of the gingiva; but still it helps to prevent food from lodging between the teeth, and it thereby protects the attachment of the gingiva to the approximal tooth surfaces.

GINGIVAL SULCUS

The *gingival sulcus* is the space, or potential space, that exists between the tooth surface and the narrow, unattached cervical margin of the gingiva. The depth of the gingival sulcus varies in different individuals, in different teeth in the same mouth, in different areas around the same tooth, and in different states of oral health. A description of the sulcus will usually be found in textbooks of oral histology.

CERVICAL LINE

This is the line of the cementoenamel junction.

Relationship of the gingival margin to the cervical line:
    a. The gingival margin approximately follows the curvature of the cervical line whether the clinical crown is short or long. Usually it is not exactly at the level of the cervical line because of the recession of the gingiva with age. In a young person the gingival margin will be on the anatomic crown of the tooth; in older persons it will often be on the anatomic root. (Examine the mouth of several persons of different ages.)

    b. The location of the *margin of the gingiva* moves apically on the tooth as the individual grows older. The location of the *cervical line* remains the same (Fig. 1-1, maxillary left canine).

TOOTH CURVATURES

1. The *location* of the crest of curvature on the facial and lingual surfaces of the crowns of teeth (Fig. 2-3) can be seen from the mesial and distal aspects to be usually in one of 2 places:

    a. In the cervical third of the crown on:

        (1) Facial and lingual surfaces of all anterior teeth (maxillary and mandibular).
        (2) Facial surface of all posterior teeth (maxillary and mandibular).
        (3) Lingual surfaces of maxillary posterior teeth (sometimes).

b. In the middle third of the crown on:

(1) Lingual surface of all mandibular posterior teeth.
(2) Lingual surface of maxillary posterior teeth (sometimes).

2. Function of facial and lingual curvatures:

The form of the curvature determines the direction of food as it is pushed cervically over the tooth surface during mastication. The curvatures are ordinarily such that they prevent trauma to the gingiva.

WHAT IS INVOLVED IN THE ACT OF CHEWING?

Food is placed between the opposing anterior teeth.

The jaws are closed in a protrusive position until the teeth meet.

The mandible then is moved posteriorly, thereby cutting a portion of the food free.

The food is transferred by the tongue to the posterior teeth; it is held in position by the cheeks and the tongue.

The teeth are brought together in a lateral position and are slid into centric occlusion.

The various incline planes, ridges, and grooves of opposing teeth reduce the food to bits. (Bite off a piece of firm food and analyze your jaw movements as you prepare to swallow it.)

Notes

Examine mouths until you see the parts and forms mentioned.

Chew on firm piece of food or on gum and analyze the action of the mandible.

13

The Maxillary and Mandibular Bones and Their Relationship to the Teeth*

The *dental arches* are attached to the upper and lower jaws. The roots of the teeth are embedded in *alveoli* (pits) in the maxillary and mandibular bones, and are attached to the bone by the fibers of the *periodontal ligament*.

Maxillae

GENERAL

The 2 maxillary bones (the right and left maxillae) form the upper jaw (Figs. 13-1 and 13-3).

The maxillae are the largest bones of the face, with the exception of the mandible. They form:

The anterior three-fourths of the hard palate.
The floor and lateral walls of the nose.
The floor of the orbit.
The cheek under the orbit, near the nose.
The walls of the maxillary sinuses.
The bone support for the teeth.

PARTS OF A MAXILLA

Each maxilla (right and left) consists of: 1 *body* and 4 *processes*.

BODY

The body is shaped somewhat like a pyramid.

It contains a large cavity—the *maxillary sinus* (Fig. 13-2).

---

*To obtain a clear understanding of these bones and their relationship to one another and to the teeth, it is almost necessary to have a skull at hand to examine while reading the outline. If you trace each bone with your fingers as you read, you are not apt to forget its characteristics.

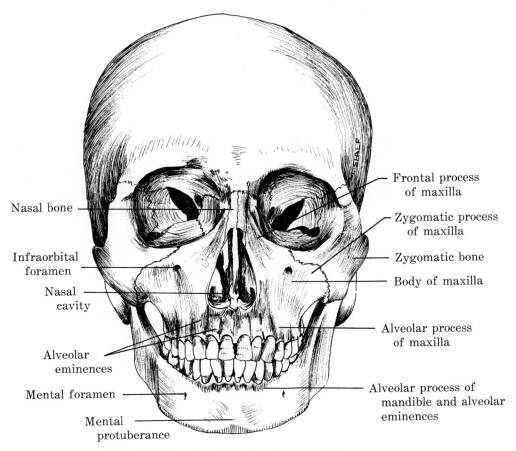

Nasal bone

Infraorbital
foramen

Nasal
cavity

Alveolar
eminences

Mental foramen

Mental
protuberance

Frontal process
of maxilla

Zygomatic process
of maxilla

Zygomatic bone

Body of maxilla

Alveolar process
of maxilla

Alveolar process of
mandible and alveolar
eminences

FIG. 13-1. Drawing of the frontal aspect of a human skull.

The infraorbital foramen is on the anterior surface of the body. It is an opening for the infraorbital nerve and vessels (Fig. 13-1).

PROCESSES

1. *Frontal process*

Its anterior edge articulates with the nasal bone.

The superior edge articulates with the frontal bone (Fig. 13-3).

The medial surface is toward the nasal cavity.

2. *Zygomatic process*

Laterally, it articulates with the zygomatic bone (cheek bone).

In front, it forms part of the anterior surface of the maxilla.

3. *Alveolar process*

This part of the maxillary bone holds the maxillary teeth (8 on either side in the permanent dentition).

4. *Palatine process* (Fig. 12-1)

This process projects medially from the inner surface of each maxillary bone and meets the palatine process of the opposite maxillary bone at the center line of the hard palate.

The inferior surfaces of the right and left palatine processes form the anterior $\frac{3}{4}$ of the hard palate.

The superior surfaces form the floor of the nasal cavity.

The palatine processes articulate posteriorly with the horizontal plates of the *palatine bones* which form the posterior $\frac{1}{4}$ of the hard palate.

HARD PALATE (Fig. 12-1)

The shape of the palate and the shape of the maxillary arch vary together.

The palate blends without a line with the palatal part of the alveolar process.

The *incisive foramen* is in the midline posterior to the central incisors.

The anteroposterior line of fusion between the palatine processes of the maxillae and between the horizontal plates of the palatine bones is visible posterior to the incisive foramen. This is the *palatine raphe* (pronounced rā'fē).

The embryonic premaxilla cannot be distinguished in the adult. It is part of the maxillary bone.

MAXILLARY SINUS (Fig. 13-2 and 14-4)

This cavity is located in the body of each maxilla, right and left.

It has the shape of a pyramid.

Sometimes only a *very* thin bone lies between the floor of the sinus and the apices of the roots of the maxillary molar teeth.

Sometimes (rarely) no bone at all separates the root apices from the sinus. There is always soft tissue, made up of the periodontal ligament on the tooth root and the sinus mucous membrane lining the sinus cavity, between the cementum of the root and the space of the cavity.

Mandible

The mandible is a single horseshoe-shaped bone, and generally speaking, it is bilaterally symmetrical. It bears the mandibular teeth and is attached by ligaments and muscles to immovable bones of the skull. The mandibular joints are movable articulations—the only movable articulations in the head.

PARTS OF THE MANDIBLE (Figs. 13-1 and 13-3)

The mandible has 3 parts: 1 *body* and 2 *rami* (singular, ramus). The body and ramus join at the *angle of the mandible* on either side.

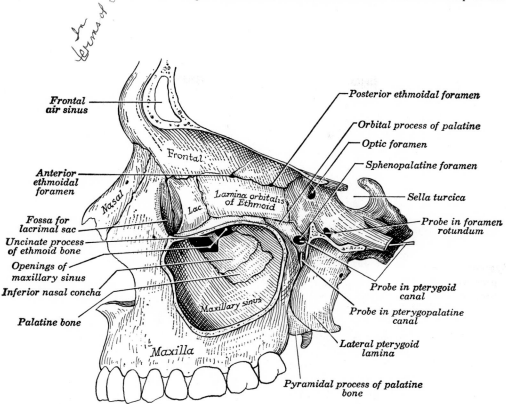

Fig. 13-2. The external wall of the body of the left maxillary bone has been removed and the large maxillary sinus is exposed. The floor of the sinus is above the maxillary posterior teeth, but it does not extend as far forward as the maxillary anterior teeth. As you can see, the floor of the nasal cavity, not the floor of the sinus, is above the anterior teeth. Notice the openings into the nasal chamber in the medial wall of the sinus. The part of the palatine bone indicated in the illustration is not the part that makes up the posterior fourth of the hard palate; it is the vertical part of the palatine bone. The posterior part of the hard palate is the horizontal part of the palatine bone (From *Gray's Anatomy of the Human Body,* 29th Ed., C. M. Goss, Editor, p. 186, Philadelphia, Lea & Febiger, 1973).

BODY (external surface)

The *symphysis* is the line of fusion at the midline.

Mental tubercles lie on either side of the symphysis.

The *mental protuberance* is above and between the tubercles.

Alveolar eminences are elevations over tooth roots.

The *mental foramen* lies near the apex of the second molar.*

The *external oblique* line extends from the mental tubercle to the anterior border of the ramus.

*In a study of 40 skulls, the mental foramen was found to be:

Under the apex of the first premolar	never
Between the apices of the first and second premolars	40%
Directly under the second premolar	42.5%
Distal to the apex of the second premolar	17.5%

Roman-Ruiz, Luis Alipio, *The Mental Foramen: A Study of Its Positional Relationship to the Lower Incisor and Premolar Teeth.* Master's Thesis, College of Dentistry, The Ohio State University, 1970, Columbus, Ohio.

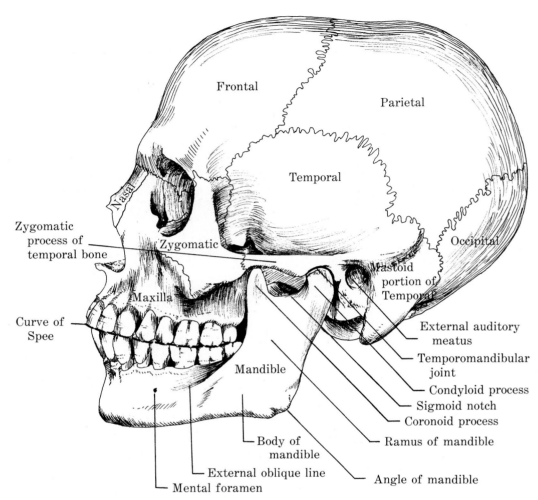

Fig. 13-3. Drawing of the left side of a human skull.

RAMUS (external surface)

There are 2 processes on the upper end of each ramus:

1. *Coronoid process*

   This process is roundly pointed on the upper border and flat on the external and internal surfaces.

   The *sigmoid notch* (mandibular notch) is between the coronoid process and the condyloid process.

2. *Condyloid process*

   This process is composed of
   A *condyle* (a head) and
   A *neck* to which the head is attached.

Shape of condyle:

From the side it looks like a round knob.

From the posterior it is long and narrow.

The condyle fits into a fossa in the temporal bone of the cranium.

INTERNAL SURFACE OF MANDIBLE

On either side of the symphysis on the internal surface of the mandible are the *genial tubercles.*

The *internal oblique line* (*mylohyoid line*) is opposite the external oblique line.

The *fossa* for the submandibular gland is below the mylohyoid line.

Mandibular foramen:

Located on the medial surface of the ramus below the sigmoid notch in the center of the ramus.

It is the entrance of inferior alveolar vessels and nerves.

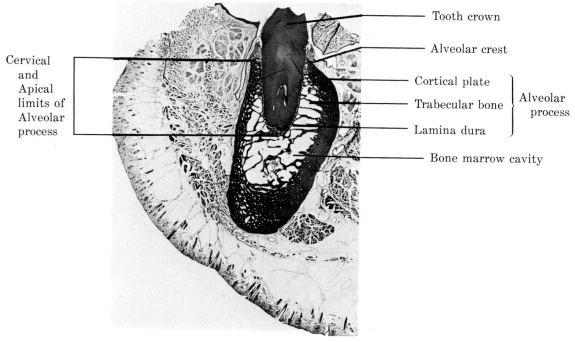

FIG. 13-4. A buccolingual section, about 30 microns thick, of a human mandible and a posterior tooth. The tooth enamel was destroyed by the decalcification of the specimen with nitric acid preparatory to embedding and sectioning. Only a small part of the pulp cavity is in this section. The cortical plate is thick, and the trabecular bone is less dense than in some specimens. The periodontal ligament is seen as a narrow white line between the tooth root and the lamina dura. The alveolar process comprises all of the bone surrounding the tooth from the alveolar crest to the root apex. To the left of the mandibular bone is the soft tissue of the cheek.

The Alveolar Process

The alveolar process is that part of the mandible or maxilla that surrounds and supports the teeth (Figs. 13-1 and 13-4).

DIVISIONS OF THE ALVEOLAR PROCESS:

1. *Lamina dura*—(synonyms: alveolar bone, true alveolar bone, alveolar bone proper, cribriform plate). The lamina dura is the bone that forms the wall of the tooth socket.

2. *Supporting bone,* made up of:

   Cortical plate—which is the outer surface, facial and lingual, of the alveolar process.

   Trabecular bone—(synonyms: cancellous bone, spongy bone). This is composed of platelike bone partitions which, inside the bone, separate bone marrow spaces of various sizes and shapes.

Notes

If you have an opportunity to examine a number of skulls, compare them for variations in bone form. Look at the following:

Shape and height of the mandible.

Location of mental foramen.

Thickness of the bone on the facial side of tooth roots.

Thickness of the bone on the palatal and lingual sides of teeth.

Relationship of the apices of maxillary incisors to the floor of the nose.

Relationship of the apices of maxillary molars to maxillary sinuses.

(For information on this see Sicher, H. and DuBrul, E. L., *Oral Anatomy*, 5th ed., St. Louis, Mosby, 1972.) This is an interesting and important anatomic relationship.

14

Nerves Associated with the Mouth

The pairs of cranial nerves which supply the mouth and the surrounding structures are:[1,2]

Trigeminal Nerve	(V)
Facial Nerve	(VII)
Glossopharyngeal Nerve	(IX)
Hypoglossal Nerve	(XII)

Trigeminal Nerve (Fifth Nerve)

The *trigeminal nerve* is the largest of the cranial nerves. It originates in the *semilunar ganglion* (*trigeminal* or *Gasserian ganglion*), which is located in a small depression on the internal surface of the *temporal bone* (Fig. 14-2).

It is a mixed nerve: it contains both afferent and efferent (sensory and motor) fibers.

AFFERENT (SENSORY) FIBERS supply the skin of the entire face, the mucous membrane of the oral and nasal cavities (excepting the pharynx and the base of the tongue), and the teeth and their supporting structures.*

EFFERENT (MOTOR) FIBERS supply the muscles of mastication and several other muscles in the region of the mouth.

THE TRIGEMINAL NERVE HAS 3 DIVISIONS (Fig. 14-1):

Division I	OPHTHALMIC NERVE	(afferent)
Division II	MAXILLARY NERVE	(afferent)
Division III	MANDIBULAR NERVE	(afferent and efferent)

Each of these 3 divisions is further divided into branches and subbranches. The branches of the *maxillary nerve* and of the *mandibular nerve* are the ones which supply the region of and about the oral cavity.

*The term *supporting structures* is used to indicate the periodontal ligament, the alveolar process, and the gingiva.

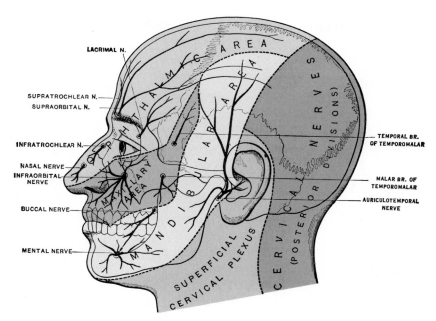

Fig. 14-1. Diagram of the general distribution of the 3 divisions of the trigeminal nerve (From *Gray's Anatomy of the Human Body*, 29th Ed., C. M. Goss, Editor, p. 918, Philadelphia, Lea & Febiger, 1973).

Ophthalmic Nerve

(DIVISION I OF THE TRIGEMINAL NERVE)

The *ophthalmic nerve* supplies the eyeball, the upper eyelid, the skin of the nose, the skin of the forehead, the skin of the scalp, the parts of the nasal mucosa, the lacrimal glands (Figs. 14-1 and 14-2).

This nerve does not supply the oral cavity.

Maxillary Nerve

(DIVISION II OF THE TRIGEMINAL NERVE)

The *maxillary nerve* has 4 principal branches (Figs. 14-2 and 14-3):

    1. SPHENOPALATINE NERVE
    2. POSTERIOR SUPERIOR ALVEOLAR NERVE
    3. INFRAORBITAL NERVE
    4. ZYGOMATIC NERVE

1. SPHENOPALATINE NERVE

This branch is closest to the origin of the *maxillary nerve*. It is attached to the spheno-palatine (pterygopalatine) ganglion (Fig. 14-3). The fibers of the sphenopalatine nerve pass through the ganglion and branches are given off. Some of these branches are:

a. NASOPALATINE NERVE OF SCARPA

This nerve passes through the incisive canal and goes to:

The palatal mucosa in the anterior region of the palate.

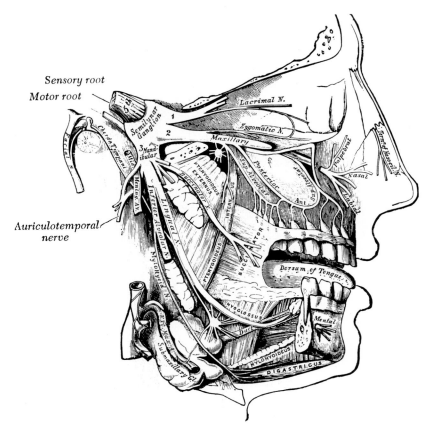

FIG. 14-2. The distribution of the maxillary and mandibular divisions of the trigeminal nerve (From *Gray's Anatomy of the Human Body*, 29th Ed., C. M. Goss, Editor, p. 916, Philadelphia, Lea & Febiger, 1973).

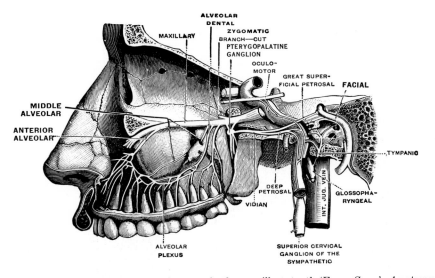

FIG. 14-3. The branches of the maxillary nerve that supply the maxillary teeth (From *Gray's Anatomy of the Human Body*, 28th Ed., C. M. Goss, Editor, p. 1019, Philadelphia, Lea & Febiger, 1966).

The palatal gingiva* of the maxillary incisor and canine teeth.

The periodontal ligament and palatal portion of the alveolar process of these teeth.

b. PALATINE NERVES

(1) *Anterior palatine nerve (Greater palatine nerve)*

After entering the oral cavity through the greater palatine foramen this nerve goes to:

The mucosa of the hard palate.

The palatal gingiva of premolar and molar teeth.

The periodontal ligament and the palatal portion of the alveolar process of these teeth.

(2) *Middle* and *Posterior palatine nerves*

After entering the oral cavity through the lesser palatine foramen these nerves go to:

Tonsils and mucosa of the soft palate.

2. POSTERIOR SUPERIOR ALVEOLAR NERVE (Figs. 14-2, 14-3)

This nerve arises from the *maxillary nerve* in 1 or 2 branches just before the *maxillary nerve* enters the infraorbital groove. Branches of the *posterior superior alveolar nerve* supply:

The *pulp of the maxillary molar teeth* through the apical foramina, with the exception of the mesiobuccal root of the maxillary first molar.

The *buccal gingiva in the maxillary molar and premolar regions.*

The periodontal ligament and the buccal portion of the alveolar process of the maxillary molar teeth.

The mucosa of the maxillary sinus and the cheek.

3. INFRAORBITAL NERVE

This name is given to the nerve which is the extension of the *maxillary nerve* after the latter has entered the infraorbital canal. It has several branches:

---

* The gingiva on the facial side of the mandible and maxillae in the premolar and molar regions is called the *buccal gingiva.*

The gingiva on the facial side of the mandible and maxillae in the incisor regions is called the *labial gingiva.*

The gingiva on the inside of the maxillary arch is called the *palatal gingiva.* This term should not be applied to the mucous membrane of the hard and soft palates. The mucous membrane in those areas is not gingiva: it is called *palatal mucosa.*

The gingiva on the inside of the mandibular arch is called the *lingual gingiva.*

a. MIDDLE SUPERIOR ALVEOLAR NERVE* (Figs. 14-2, 14-3)

This branch arises from the *infraorbital nerve* and supplies:

The *pulp of the maxillary premolar teeth* through the apical foramina, and the *pulp of the maxillary first molar* through the foramen of the mesiobuccal root.

The periodontal ligament and the buccal portion of the alveolar process in the maxillary premolar region.

The mucosa of the maxillary sinus.

b. ANTERIOR SUPERIOR ALVEOLAR NERVE (Figs. 14-2, 14-3)

This branch arises from the *infraorbital nerve* and supplies:

The pulp of the maxillary incisor and canine teeth through the apical foramina.

The *labial gingiva in the maxillary incisor and canine region.*

The periodontal ligament and the labial portion of the alveolar process in the same region.

The mucosa of the maxillary sinus and of the nasal cavity.

c. TERMINAL BRANCHES OF THE INFRAORBITAL NERVE

These branches supply:

The mucosa of the upper lip.

The skin of the upper lip, lower eyelid, and nose.

4. ZYGOMATIC NERVE

This nerve supplies the bone and skin on the prominence of the upper cheek and temporal region, and the orbit.

Mandibular Nerve

(DIVISION III OF THE TRIGEMINAL NERVE)

The *mandibular nerve* is a mixed nerve: that is, it contains both afferent and efferent fibers. It contains the only efferent portion of the fifth nerve.

---

*Comparison of descriptions of the superior alveolar nerves indicates a great lack of uniformity in their distribution. Sometimes the middle superior alveolar nerve is missing and the function is taken over by the anterior and posterior alveolar nerves.

EFFERENT (*Motor*) *Nerves* of the Mandibular Nerve:

Those nerves supplying the muscles of mastication (Figs. 16-4a, b; 16-5; 16-6):

1. MASSETERIC NERVE—to masseter muscle.

2. POSTERIOR and ANTERIOR TEMPORAL NERVE—to temporalis muscle.

3. MEDIAL PTERYGOID NERVE—to medial pterygoid muscle.

4. LATERAL PTERYGOID NERVE—to lateral pterygoid muscle.

Those nerves supplying other muscles in the region of the oral cavity:

1. MYLOHYOID NERVE (mylohyoid muscle and anterior belly of the digastric muscle).

2. TENSOR VELI PALATINI (1 muscle of the soft palate).

AFFERENT (*Sensory*) *Nerves* of the Mandibular nerve (There are 4 Divisions):

1. LONG BUCCAL NERVE (Buccinator nerve: Fig. 14-2)

   This nerve supplies:

   The *buccal gingiva in the area of the mandibular molars,* and sometimes in the area of the premolars.

   The periodontal ligament and buccal portion of the alveolar process in the molar, and sometimes in the premolar area.

   The mucosa of the cheek and corners of the mouth.

   The skin of the cheek and corners of the mouth.

2. LINGUAL NERVE (Fig. 14-2)

   This nerve supplies:

   The *lingual gingiva of the entire mandibular arch.*

   The periodontal ligament and the lingual portion of the alveolar process of all mandibular teeth.

   The mucosa on the inner surface of the mandible and in the sublingual region.

   The mucosa on the upper and lower surface of the body of the tongue, i.e., the area including and anterior to the circumvallate papillae. *This nerve supplies only general sensation:* touch, pain, pressure, temperature. It does not supply the sense of taste. *The sense of taste in this region is supplied by the facial nerve (VII).*

3. INFERIOR ALVEOLAR NERVE (Fig. 14-2)

This is actually a mixed nerve, but it is mostly sensory (afferent). Its branches are:

a. *Mylohyoid nerve*

This branch of the *inferior alveolar nerve* is the efferent (motor) portion of the *inferior alveolar nerve*. It supplies the mylohyoid muscle and the anterior belly of the digastric muscle.

b. *Dental branches of the inferior alveolar nerve* (Fig. 14-4)

These nerves supply:

The *pulp of all of the mandibular teeth* through their apical foramina.

The *facial and lingual gingiva of all mandibular teeth.*

The periodontal ligament and the facial and lingual portions of the alveolar process of all mandibular teeth.

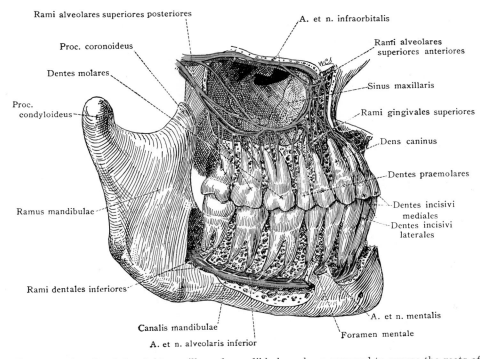

FIG. 14-4.   The external walls of the right maxilla and mandible have been removed to expose the roots of the teeth. Also, the external wall of the body of the maxillary bone has been removed to expose the maxillary sinus. (Examine particularly the relationship of the root tips of the maxillary posterior teeth to the floor of the sinus.) In this picture the nerves are shown in yellow, the arteries in red (From *Gray's Anatomy of the Human Body,* 29th Ed., C. M. Goss, Editor, p. 1170, Philadelphia, Lea & Febiger, 1973).

c. *Mental nerve* (Fig. 14-2)

This nerve supplies:

The *labial gingiva of the mandibular incisors and canines.*

The periodontal ligament and the facial portion of the alveolar process of these teeth.

Sometimes the *buccal gingiva of the premolars.*

The mucosa and skin of the lower lip.

4. AURICULOTEMPORAL NERVE (Fig. 14-2)

This nerve supplies:

Temporomandibular joint.

Outer ear.

Skin of the posterior parts of the temple and cheek.

Parotid gland.

Facial Nerve (Seventh Nerve)

The *facial nerve* is a mixed nerve (efferent and afferent).

EFFERENT (*Motor*) *Fibers* of the *Facial Nerve:*

These innervate:

Muscles of expression of the face and the scalp.

Posterior belly of the digastric muscle.

SECRETORY *Fibers* of the *Facial Nerve:*

These are efferent fibers, but cannot be called "motor":

To sublingual and submandibular (submaxillary) glands, effecting secretion.

AFFERENT (*Sensory*) *Fibers* of the *Facial Nerve:*

The *chorda tympani* branch supplies the *sense of taste to the anterior two-thirds of the tongue* (the body and the tip of the tongue).

Glossopharyngeal Nerve (Ninth Nerve)

This is a mixed nerve (efferent and afferent).

It supplies parts of the tongue and pharynx.

EFFERENT (*motor*) *fibers* of the *Glossopharyngeal Nerve* innervate a muscle of the pharynx (*stylopharyngeus*).

SECRETORY *fibers* of the *Glossopharyngeal Nerve* cannot be called "motor"; they innervate the *parotid gland*, effecting secretion.

AFFERENT (*Sensory*) *Fibers* of the *Glossopharyngeal Nerve:*

These supply the *sense of taste to the posterior one-third of the tongue*, and general sensation of the same area.

Innervate the mucosa of the pharynx and tonsils.

Hypoglossal Nerve (Twelfth Nerve)

EFFERENT (*MOTOR*) *NERVE*

This nerve supplies the muscles of the tongue.

Summary

NERVE SUPPLY TO THE TEETH

MAXILLARY TEETH

MOLARS

Posterior superior alveolar nerve (excepting the mesiobuccal root of first molar).

PREMOLARS

Middle superior alveolar nerve (supplies also mesiobuccal root of first molar).

ANTERIOR TEETH

Anterior superior alveolar nerve

MANDIBULAR TEETH

All mandibular teeth are supplied by the *dental branches of the inferior alveolar nerve.*

NERVE SUPPLY TO THE SUPPORTING STRUCTURES (GINGIVA, ALVEOLAR PROCESS, AND PERIODONTAL LIGAMENT)

TO SUPPORTING STRUCTURES OF THE MAXILLA

FACIAL SIDE OF ARCH

In molar and premolar regions:
Posterior superior alveolar nerve.
Middle superior alveolar nerve.

In incisor and canine regions:
Anterior superior alveolar nerve.

PALATAL SIDE OF ARCH

In molar and premolar regions:
Anterior palatine nerve (Greater palatine)

In incisor and canine regions:
Nasopalatine nerve of Scarpa

TO SUPPORTING STRUCTURES OF THE MANDIBLE

FACIAL SIDE OF ARCH

In molar and premolar areas:
Buccal nerve
Inferior alveolar nerve
In premolar area sometimes also *mental nerve.*

In incisor and canine areas:
Mental nerve
Inferior alveolar nerve

LINGUAL SIDE OF ARCH

Lingual nerve in all areas.
Inferior alveolar nerve in all areas.

NERVE SUPPLY TO PALATAL MUCOSA

HARD PALATE

Anterior palatine nerve
Nasopalatine nerve (in anterior region)

SOFT PALATE

Middle and posterior palatine nerves
Mandibular nerve (to *tensor veli palatini*)

NERVE SUPPLY TO THE TONGUE

BODY OF THE TONGUE

AFFERENT (SENSORY) FIBERS

Lingual nerve—general sensation   (Division III of Fifth Nerve)
Facial nerve—taste sensation   (Seventh nerve)

EFFERENT (MOTOR FIBERS)

Hypoglossal nerve   (Twelfth nerve)

BASE OF THE TONGUE

AFFERENT (SENSORY) FIBERS

Glossopharyngeal nerve—taste sensation and general sensation (Ninth Nerve)

EFFERENT (MOTOR) FIBERS

Hypoglossal nerve (Twelfth Nerve)

References

1. Edwards, L. F. and Gaughran, G. R. L. *Concise Anatomy,* 3rd ed., McGraw-Hill Book Co., New York, 1971, pp. 299–310.
2. Sicher, H. and DuBrul, E. L. *Oral Anatomy,* 5th ed., Mosby, St. Louis, 1970, pp. 344–378.

Notes

15

Blood Vessels Associated
with the Mouth

Arteries[1,2]

COMMON CAROTID (right and left) ascends in the neck and divides into:

1. EXTERNAL CAROTID (Fig. 15-1)

2. INTERNAL CAROTID (Enters skull. Does not supply the mouth.)

EXTERNAL CAROTID—As it passes up the neck, it gives off the following branches which supply the mouth area. (Branches that supply areas other than the mouth are not listed.)

1. LINGUAL ARTERY—Supplies the tongue.

2. EXTERNAL MAXILLARY ARTERY (facial artery)—Passes under the lower border of the mandible, upward over the outer surface of the mandible, and branches into:

   a. *Ascending palatine artery*, which sends branches to:

      Soft palate
      Pharyngeal muscles
      Mucosa of pharynx
      Palatine tonsil

   b. *Submental artery*, which sends branches to:

      Anterior belly of digastric muscle

      Mylohyoid muscle

   c. *Inferior labial artery*

   d. *Superior labial artery*

   e. *Artery to nose*

   f. *Artery to muscles around mouth*

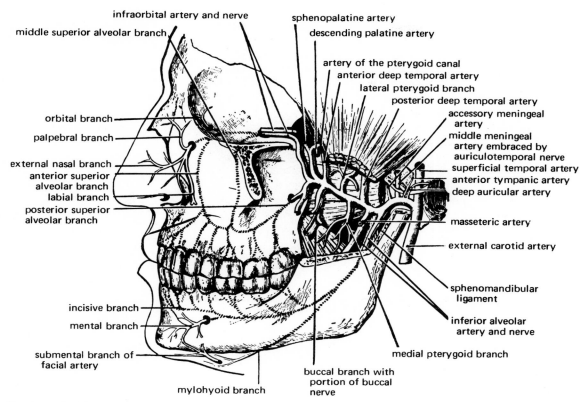

infraorbital artery and nerve

middle superior alveolar branch

sphenopalatine artery

descending palatine artery

artery of the pterygoid canal
anterior deep temporal artery
lateral pterygoid branch
posterior deep temporal artery
accessory meningeal artery
middle meningeal artery embraced by auriculotemporal nerve
superficial temporal artery
anterior tympanic artery
deep auricular artery

orbital branch

palpebral branch

external nasal branch
anterior superior alveolar branch
labial branch
posterior superior alveolar branch

masseteric artery

external carotid artery

sphenomandibular ligament

inferior alveolar artery and nerve

incisive branch

mental branch

submental branch of facial artery

medial pterygoid branch

mylohyoid branch

buccal branch with portion of buccal nerve

FIG. 15-1. Left maxillary artery. Trace the paths of the posterior superior alveolar, middle superior alveolar, and anterior superior alveolar arteries to the teeth supplied by them. Trace the inferior alveolar artery and nerve to the teeth supplied by them. Trace the inferior alveolar artery and nerve to the teeth supplied, and notice the mental branch leaving the mandible through the mental foramen (From *Concise Anatomy*, L. F. Edwards and Gr. L. Gaughran, 3rd Ed., p. 331, New York, McGraw-Hill, 1971).

3. INTERNAL MAXILLARY ARTERY—Arises from the external carotid within the parotid gland. The branches of this artery which are directly concerned with the blood supply to the teeth are (Fig. 15-1):

a. *Inferior alveolar artery,* which enters the mandible through the mandibular foramen (Figs. 15-1 and 14-4).

(1) Gives off branches to the mandibular molar teeth.

(2) Then divides into:

(a) *Mental branch,* which exists through the mental foramen onto the outside of the mandible.

(b) *Incisive branch,* which supplies the anterior teeth and the bone of this area.

b. *Posterior superior alveolar artery,* which supplies:

Maxillary molars, and
Lining of maxillary sinus

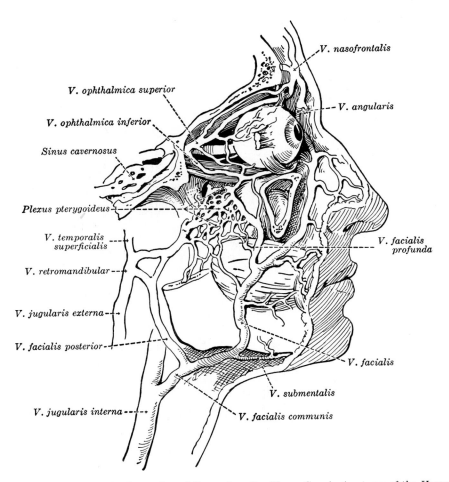

V. nasofrontalis

V. ophthalmica superior

V. ophthalmica inferior

Sinus cavernosus

V. angularis

Plexus pterygoideus

V. temporalis superficialis

V. facialis profunda

V. retromandibular

V. jugularis externa

V. facialis posterior

V. facialis

V. jugularis interna

V. submentalis

V. facialis communis

Fig. 15-2.   Principal veins that drain the region of the oral cavity (From *Gray's Anatomy of the Human Body*, 29th Ed., C. M. Goss, Editor, p. 686, Philadelphia, Lea & Febiger, 1973).

    c. *Infraorbital artery* (Fig. 14-4), which gives off:

        (1) *Anterior and Middle Superior Alveolar Arteries*, which supply

           Maxillary premolar teeth
           Maxillary canine teeth
           Maxillary incisor teeth

        (2) Branches to cheek, root of upper lip, and area of eye.

Veins[1,2]

Drainage takes place through (Fig. 15-2):

1. PTERYGOID PLEXUS, which is a plexus of veins medial to the upper part of the ramus of the mandible.
        This plexus receives blood from the area of the upper part of the face, the lips and muscles around the mouth, the palate, the maxillary alveolar process and the maxillary teeth.

2. ANTERIOR FACIAL VEIN, which receives blood from the superior and inferior labial veins, and from the muscles of mastication.

3. INFERIOR ALVEOLAR VEIN, which drains the mandible and the mandibular teeth.

4. LINGUAL VEINS, which drain the tongue.

All empty indirectly into the INTERNAL JUGULAR VEIN.

References

1. Edwards, L. F. and Gaughran, G. R. L., *Concise Anatomy*, 3rd ed., McGraw-Hill Book Co., New York, 1971.
2. Sicher, H. and Du Brul, E. L., *Oral Anatomy*, 5th ed., Mosby, St. Louis, 1970.

Notes

16

The Temporomandibular Joint

Definition

A *JOINT*, or articulation, is a connection between 2 separate parts of the skeleton (Fig. 16-1).

TEMPOROMANDIBULAR JOINT. This is the articulation between the *mandible* and the *temporal bone* of the skull. It is a bilateral articulation: the right and left sides work as a unit. It is the only free-moving articulation in the head; all others are sutures and are immovable.

PALPATE THE

TEMPOROMANDIBULAR JOINT

In front of the ear. Put your finger immediately in front of either ear and open and close your mouth.

In the ear opening. Put your finger inside of either ear, open and close your mouth, and move the mandible from side to side.

Construction of the Temporomandibular Joint

THERE ARE THREE ARTICULATING PARTS TO THE TEMPOROMANDIBU-LAR JOINT (Fig. 16-2)

MANDIBULAR CONDYLE
ARTICULAR FOSSA AND ARTICULAR EMINENCE OF THE TEMPORAL BONE
DISC

These parts are enclosed by a *fibrous capsule* (Fig. 16-3).

Examine a skull and study how the mandibular condyle fits into the articular fossa. The disc is not present in a prepared skull because the disc is not bone; and for the same reason, the fibrous capsule is not present.

MANDIBULAR CONDYLE

This articulating surface is located on the superior and anterior surfaces of the head of the condyloid process of the mandible.

153

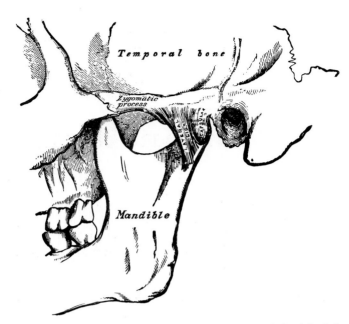

FIG. 16-1. Lateral aspect of mandible and part of cranium. The temporomandibular joint is here enclosed in the fibrous capsule, reinforced by the temporomandibular ligament (lateral ligament) (From *Gray's Anatomy of the Human Body*, 29th Ed., C. M. Goss, Editor, p. 293, Philadelphia, Lea & Febiger, 1973).

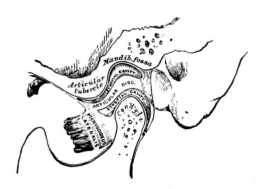

FIG. 16-2 A sagittal section through the joint shown in figure 16-1 reveals the condyle of the mandible, the articular fossa (the anterior part of the mandibular fossa), and the disc. The articular eminence is a ridge extending medially from the articular tubercle. You will see this eminence if you examine the fossa on the undersurface of the skull (From *Gray's Anatomy of the Human Body*, 29th Ed., C. M. Goss, Editor, p. 293, Philadelphia, Lea & Febiger, 1973).

FIG. 16-3. Medial aspect of the mandible and the temporomandibular joint. The fibrous capsule (capsular ligament) encloses the joint. Examine your prepared skull and locate the attachments of the stylomandibular and sphenomandibular ligaments. Clearly, these ligaments function to prevent non-functional, damaging movements in the joint (From *Gray's Anatomy of the Human Body*, 29th Ed., C. M. Goss, Editor, p. 293, Philadelphia, Lea & Febiger, 1973).

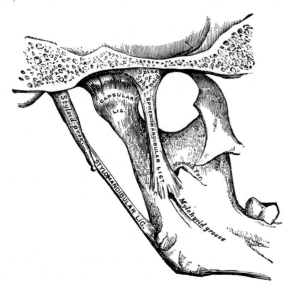

This surface is made up of fibrous connective tissue which contains a number of cartilage cells.

ARTICULAR FOSSA AND ARTICULAR EMINENCE

The articular fossa is the anterior part of the mandibular fossa.

The articular eminence is located just anterior to the articular fossa.

DISC

The disc is a body of dense fibrous connective tissue located between (1) the mandibular condyle and (2) the articular fossa and articular eminence (Fig. 16-2).

It is a thin, oval plate, thinner in the center than around the edges. It acts as a buffer between the temporal bone and the mandibular condyle.

The upper surface of the disc is concave anteroposteriorly and convex mediolaterally, thus conforming to the shape of the articular eminence.

The lower surface of the disc is concave in both directions, thus adapting to the mandibular condyle.

FIBROUS CAPSULE

The fibrous capsule is a sheet of tissue, or a sac of tissue, that encloses the joint like a tube (Figs. 16-1 and 16-3).

The upper border of the capsule is attached to the temporal bone around the circumference of the articulating surface, i.e., around the edge of the articular fossa and the articular eminence.

Its lower border is attached around the neck of the condyloid process, thus enclosing the condyle.

The fibrous capsule is composed of 2 layers:

1. *Inner layer* (synovial membrane). This is a thin layer of tissue that secretes a fluid, *synovia*, which lubricates the joint.

2. *Outer layer*. This is a layer of fibrous tissue that is reinforced by accessory ligaments which strengthen it.

RELATIONSHIP OF CAPSULE TO DISC AND OF DISC TO CONDYLE

Anteriorly the disc and the capsule are fused.

Posteriorly the disc and the capsule are connected by a thick pad of loose, vascular connective tissue, which gives the disc freedom to move anteriorly.

Laterally and *medially* the disc is attached to the lateral and medial sides of the mandibular condyle. It is not attached to the capsule laterally and medially.

The attachment of the disc to the lateral and medial borders of the mandibular condyle results in the simultaneous anteroposterior movement of the condyle and the disc—that is, the disc follows the movement of the condyle when the muscles move the mandible forward or backward; and this kind of attachment results also in a swinging movement of the mandible at the junction of the condyle and the disc.

LIGAMENTS WHICH SUPPORT THE JOINT (Figs. 16-1, 16-3)

(Ligaments do not move the joint. Muscles move the joint.)

1. The *temporomandibular ligament* is the strong reinforcement of the *lateral* wall of the capsule. It helps to prevent displacement of the mandible (Fig. 16-1).

   It is attached above to the zygomatic arch, and below to the lateral surface and the posterior border of the neck of the condyle.

2. The *stylomandibular ligament* is *posterior* to the joint and is separated from it, but it gives support to the mandible (Fig. 16-3).

   It is attached above to the styloid process of the temporal bone, and below to the posterior border and angle of the mandible.

3. The *sphenomandibular ligament* is *medial* to the joint and gives support to the mandible. It makes a strong anchor between the mandible and the skull (Fig. 16-3).

   It is attached above to the angular spine of the sphenoid bone, and below to the inside surface of the mandible at the lower border of the mandibular foramen and at the lower border of the groove of the mandibular neck.

NERVE SUPPLY TO THE TEMPOROMANDIBULAR JOINT

The temporomandibular joint is innervated by the *auriculotemporal branch* of the *third division of the trigeminal nerve* (Fig. 14-2).

BLOOD SUPPLY TO THE TEMPOROMANDIBULAR JOINT

The blood supply comes from a branch of the *superficial temporal artery*, which is from the external carotid artery.

Functioning of the Temporomandibular Joint

MUSCLES OF MASTICATION move the joint. These muscles are *paired:* right and left.

Masseter Muscle
Temporal Muscle
Medial Pterygoid Muscle
Lateral Pterygoid Muscle

1. MASSETER MUSCLE (Fig. 16-4a, b)

   It is the most superficial of the muscles of mastication.

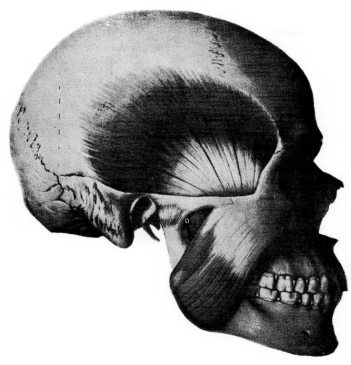

FIG. 16-4a. Masseter and temporal muscles (From *Anatomie für Zahnärzte*, H. Sicher and J. Tandler, Vienna, Julius Springer, 1928).

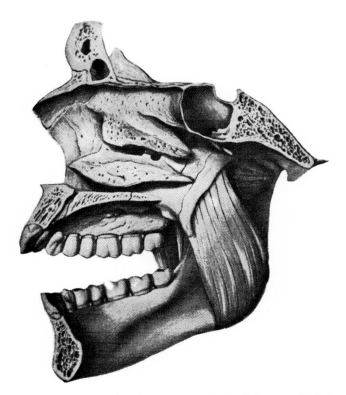

FIG. 16-4b. Medial pterygoid muscle, medial view (From *Anatomie für Zahnärtze*, H. Sicher and J. Tandler, Vienna, Julius Springer, 1928).

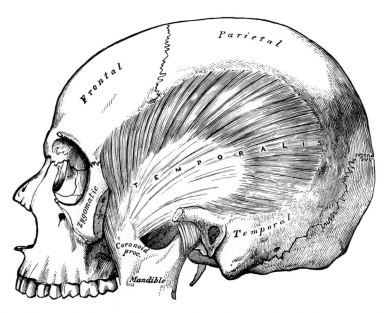

FIG. 16-5. Contraction of the anterior fibers of the temporal muscle acts to close the jaw; contraction of the posteriorly placed fibers acts to pull the jaw back (From *Gray's Anatomy of the Human Body*, 29th Ed., C. M. Goss, Editor, p. 388, Philadelphia, Lea & Febiger, 1973).

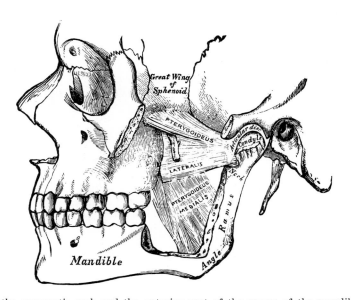

FIG. 16-6. Removal of the zygomatic arch and the anterior part of the ramus of the mandible reveals the medial and lateral pterygoid muscles. Contraction of the medial pterygoid muscle elevates (closes) the mandible. Contraction of the lateral pterygoid pulls the condyle and disc forward, which action causes the mouth to open (From *Gray's Anatomy of the Human Body*, 29th Ed., C. M. Goss, Editor, p. 388, Philadelphia, Lea & Febiger, 1973).

Origin: Zygomatic arch.*

Insertion: Lateral surface of the coronoid process, the ramus, and the angle of the mandible.

Direction: Diagonally down and posteriorly.

Action: Elevator (closes jaw)—superficial portion; and retractor (pulls jaw back)—deep portion.†

2. TEMPORAL MUSCLE (Figs. 16-4a and 16-5)

It is a fan-shaped muscle.

Origin: Entire temporal fossa.

Insertion: Coronoid process of the mandible and the anterior border of the ramus of the mandible.

Direction: Downward (anterior part), and downward and anteriorly (posterior part). The fibers pass medial to the zygomatic arch.

Action: Elevator (closes jaw); and retractor (pulls jaw back). The posterior portion retracts the mandible.

3. MEDIAL PTERYGOID MUSCLE (Figs. 16-4b and 16-6)

It is located medial to the ramus of the mandible.

Origin: It has two heads. Mainly from the median surface of the lateral pterygoid plate of the sphenoid bone. Also from the pyramidal process of the palatine bone.

Insertion: Median surface of the mandible just above the angle.

Direction: Downward and laterally.

Action: Elevator (closes jaw).

4. LATERAL PTERYGOID MUSCLE (Fig. 16-6)

It is located deep in the infratemporal fossa.

Origin: It has 2 heads: at 2 places on sphenoid bone.

Insertion: Front of the neck of the condyloid process and to the anteromedial surface of the capsular ligament (and indirectly, thereby, to the disc).

Direction: Posteriorly and laterally.

Action: Opens jaws. It does this by pulling the articular disc and the condyle forward. Assisted by geniohyoid and digastric muscles.

Protrudes mandible.

*There may be several areas of attachment of these various muscles; not all such areas are named.

†There may be other types of movements produced by the contraction of these muscles; not all such movements are named.

Lateral movement of the mandible is also produced by these muscles:
Contraction of the left lateral pterygoid muscle produces movement to the right.

Contraction of the right lateral pterygoid muscle produces lateral movement to the left.

ELEVATION OF MANDIBLE (closing of mouth) results from the contraction of: *

Masseter Muscle
Temporalis Muscle
Medial Pterygoid Muscle

DEPRESSION OF MANDIBLE (opening of mouth) results from the contraction of:

Lateral Pterygoid Muscle, assisted by
Anterior Belly of Digastric Muscle
Geniohyoid Muscle

MOVEMENT OF TEMPOROMANDIBULAR JOINT

The disc divides the articular space into upper and lower compartments. The result of this division is the formation of two joints.

Upper joint: The disc slides forward and backward over the articular eminence, resulting in protrusion and retraction during opening and closing of the mouth. Such sliding motion, when it occurs unilaterally on alternate sides, results in a *grinding motion.* (Place your fingers in front of your ears and open and close your mouth.)

Lower joint: In the lower joint there is only a hinge motion. If the upper joint is immobilized, there is no protrusion, retraction, or grinding during opening and closing: there is only a swinging of the mandible at the junction of the disc and the mandibular condyle.

In human beings, automatic opening is a combination of a hinge and sliding movement. This seems to be a specifically human characteristic. Even in the highest order of apes (chimpanzee) the mandible drops in a simple hinge movement.

DISLOCATION OF THE MANDIBLE

During an extreme opening of the mandible, the disc may slip out of the articular fossa and forward over the articular eminence, and the mandible will be dislocated.

This dislocation occurs in the upper joint: you recall that the disc is attached to the condyle and travels with it.

Because of its looseness, the capsule does not tear when dislocation occurs.

*Other muscles are involved in the complex mandibular movements.

Development of the Temporomandibular Joint

IN INFANTS. The glenoid fossa, the articular eminence, and the condyle are rather flat. This allows for a wide range of sliding motions in the temporomandibular joint.

DURING DEVELOPMENT. The glenoid fossa deepens, the articular eminence becomes prominent, the condyle becomes rounded, and the shape of the disc changes to conform to the change in shape of the fossa and the condyle.

GROWTH. The condyle contains cartilage beneath its surface, and the condyloid process grows in length until the age of 20 to 25 years. This is one way the mandible grows in depth. As a result of growth in the condyle area, the body of the mandible is lowered from the skull.

References

1. Edwards, L. F. and Gaughran, G. R. L., *Concise Anatomy*, 3rd ed., McGraw-Hill Book Co., New York, 1971.
2. Sicher, H. and DuBrul, E. L., *Oral Anatomy*, 5th ed., Mosby, St. Louis, 1970.

Notes

Study a skull and see how the mandibular condyle fits into the mandibular fossa. The condyle, with the disc superimposed and attached medially and laterally, functions in the anterior part of this fossa. (The anterior part of the mandibular fossa is called the articular fossa.)

Feel the function of the temporomandibular joint by placing a finger on either side of your face just in front of your ears and opening and closing your mouth and moving the mandible from side to side. Then put your fingers in your ears and move the mandible.

Now try to reproduce this action with the skeleton. (Of course, the skeleton has lost the disc, which is not bone.) If you are moving the mandible correctly the condyle will move down onto the articular eminence. (See Sicher and DuBrul, which is cited in the reference section, for interesting information.)

17

Occlusion*

Definition

Occlusion is contact of the masticating and incising surfaces of maxillary and mandibular teeth. To *occlude* means *to close*.

In dentistry the study of occlusion includes study of the anatomy, physiology, and pathology of the teeth, bones, and soft tissues of the oral cavity during function.

Understanding occlusion requires knowledge of:
1. The arrangement of teeth in the dental arches.

2. The relation of the mandibular dental arch to the maxillary dental arch.

3. The relation of the mandible to the maxilla.

4. The temporomandibular joint.

5. The muscles, nerves, ligaments, and soft tissues that affect the position of the mandible.

6. Abnormalities that may be detrimental to dental health.

Normal Arrangement of Teeth in the Dental Arches†

When viewed from the occlusal aspect, each dental arch is U-shaped.

The incisal edges and buccal cusp tips follow a curved line around the outer edge of the dental arch; the lingual cusp tips of the posterior teeth follow a curved line nearly parallel to the buccal cusp tips.

Curve of Spee—When viewed from the buccal aspect, the cusp tips of posterior teeth follow a gradual curve anteroposteriorly. The curve of the maxillary arch is convex, of the mandibular arch, concave (Fig. 13-3).

---

*This chapter was contributed by Gary L. Racey, D.D.S., Instructor in Oral Surgery, The Ohio State University, Columbus, Ohio.

†Examine mouths of your associates and look carefully at all of the characteristics described. Also examine your prepared skulls.

Curve of Wilson—When viewed from the anterior aspect with the mouth slightly open, the cusp tips of the posterior teeth follow a gradual curve from the left side to the right side. The curve of the maxillary arch is convex, of the mandibular arch concave.

Normal Relation of the Mandibular Dental Arch to the Maxillary Dental Arch

When the maxillary and mandibular dental arches are closed in a normal fashion, the following relationships are seen:

1. THE MAXILLARY TEETH overlap the mandibular teeth.

   a. *Horizontal overlap*—The incisal edges of maxillary anterior teeth are labial to the incisal edges of the mandibular teeth.

   b. *Vertical overlap*—The incisal edges of the maxillary anterior teeth extend below the incisal edges of the mandibular teeth.

   c. The maxillary posterior teeth are slightly buccal to the mandibular posterior teeth so that

      1) the buccal cusps of the maxillary teeth are buccal to the mandibular arch.

      2) the lingual cusps of maxillary teeth rest in the occlusal fossae of the mandibular arch.

      3) the buccal cusps of the mandibular teeth rest in the occlusal fossae of the maxillary arch.

      4) the lingual cusps of the mandibular teeth are lingual to the maxillary arch.

2. THE VERTICAL (LONG) AXIS OF EACH MAXILLARY TOOTH is slightly distal to the vertical axis of the corresponding mandibular tooth. Cusps of the maxillary teeth rest between cusps of the mandibular teeth, rather than cusp tip on cusp tip. The fitting together of cusps in occlusion is called *intercuspation*.

3. EACH TOOTH IN A DENTAL ARCH occludes with 2 teeth in the opposing arch except the mandibular central incisor (which is narrower than the maxillary central incisor) and the maxillary last molar.

Examine several mouths and look carefully at the arrangement and relationship of the teeth with the dental arches separated and with the teeth in occlusion.

Jaw Relations—the Relationship of the Mandible to the Maxilla*

Jaw relation refers to the position of the mandible relative to the maxilla and must be thought of as *bone to bone relationship*, not as tooth to tooth relationship.

---

*When teeth are not present, vertical and horizontal jaw relations still exist. Dentists who make dentures attempt to duplicate the patient's jaw relationships when teeth were present. An articulator is a mechanical device which helps him to study and duplicate the patient's jaw relations.

The maxilla is firmly attached at suture lines to other bones of the facial skeleton: its position is fixed.

The mandible articulates with the skull at the temporomandibular joint: it can move relative to the maxilla.

For convenience, the relations of the mandible to the maxilla are described as *vertical relation* and *horizontal relation.*

1. VERTICAL RELATION refers to the amount of separation between the mandible and maxilla.

   a. *Vertical relation of occlusion*

      This is the amount of separation between the mandible and maxilla when the teeth (when present) are in natural occlusion.

   b. *Vertical relation of rest position*

      In a normal, erect posture when no conscious effort is made to close the teeth, there is a 2 to 6 mm. (average) space between the occlusal surfaces of the maxillary and mandibular dental arches. Vertical relation of the rest position is the amount of separation between the mandible and maxilla when the teeth (when present) are in this resting position. The 2 to 6 mm. distance between the maxillary and mandibular dental arches is called the *interocclusal distance.*

2. HORIZONTAL RELATION refers to the anteroposterior and the lateral positions of the mandible relative to the maxilla.

   a. *Centric relation*

      The most posterior position of the mandible relative to the maxillae. This is a relationship of the bones of the upper and lower jaws; the presence or absence of teeth is not a factor.

   b. *Centric occlusion*

      The maximum contact attainable between maxillary and mandibular teeth.

   c. *Centric relation occlusion*

      The contact between maxillary and mandibular teeth when the mandible is in centric occlusion.

   d. *Other horizontal relations*

      1) *Protrusive Relation*—the mandible is moved anteriorly (as when incising food) so that both mandibular condyles move forward in the mandibular fossae.

      2) *Lateral Relation*—the mandible is moved to the right or left side (as when masticating food). If the mandible moves to the right side, the right condyle remains relatively fixed and the left condyle moves forward in the mandibular fossa.

Working side—the side toward which the mandible moves.

Balancing side—the other side.

The Temporomandibular Joint

Review the section on the temporomandibular joint, pp. 153–161.

The temporomandibular joint is a ginglymo-arthroidal joint permitting rotation and sliding movements. The important movements in the joint are:

1. *Rotational Opening* (hinge-axis opening)—opening the mandible by pure rotation of the condyles in the joint. Many dentists believe that this is possible only in centric relation, and with conscious effort.

2. *Translatory Motion*—translation is bodily motion of a mass in one direction. It may follow a curved path. Most functional movements of the mandible are translatory motions with curved components.

Soft Tissues that Affect the Position of the Mandible

MUSCLES

The muscles of mastication have major control over movements of the mandible. *Review the section on muscles of mastication, pp. 156–160.*

The suprahyoid muscle group, the infrahyoid muscle group, and the posterior and deep muscles of the neck have postural influence on the mandible and maxilla.

NERVES

The *trigeminal nerve* supplies efferent (motor) neurons to the muscles of mastication.

The *trigeminal nerve* supplies afferent (sensory) neurons that provide the brain with information about the position of the teeth and jaws. The interpretation of postural information by the brain is called *proprioception*.

The periodontal ligament is well supplied with proprioceptive neurons.

The temporomandibular joint has proprioceptive neurons.

Review the section on the trigeminal nerve, pp. 136–143.

LIGAMENTS

The temporomandibular ligament, the sphenomandibular ligament, and the stylomandibular ligament are the ligaments of the temporomandibular joint.

Many people believe that in addition to supporting the temporomandibular joint, these ligaments limit movement of the joint and mandible.

CLASSIFICATION OF MALOCCLUSION

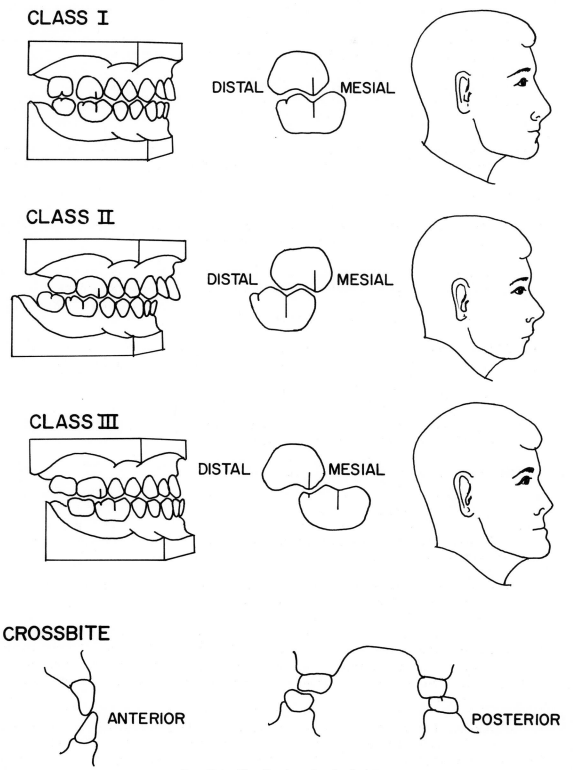

FIG. 17-1. Classification of malocclusion.

OTHER SOFT TISSUES

The cheeks (including muscles of facial expression) and the tongue are thought to influence development, position, and shape of the dental arches, the maxilla and the mandible.

Fascia is connective tissue that forms sheets or bands between anatomic structures. Fascia attaches to bones and surrounds muscles, glands, vessels, nerves, and fat. Fascia is thought to limit movement.

Malocclusion

Malocclusion results from abnormalities in the arrangement of teeth or in the occlusion of the dental arches that are detrimental to oral health.

Malocclusions are described as either *skeletal abnormalities* or as *dental abnormalities:*

1. SKELETAL MALOCCLUSION—results from abnormal development or position of the mandible and/or the maxilla. Common abnormalities are:

   a. *Mandibular prognathism*—the mandible is large relative to the maxilla. The mandibular teeth are often in cross-bite occlusion. The chin is prominent.

   See: *Class III* Malocclusion (Fig. 17-1).

   b. *Mandibular retrognathism*—the mandible is small relative to the maxilla. The mandibular teeth are often lingual to the maxillary arch. The chin appears receded.

   See: *Class II* Malocclusion (Fig. 17-1).

2. DENTAL MALOCCLUSION—results from abnormal arrangement or occlusion of the dental arches when no skeletal abnormality exists.

   See: *Class I* Malocclusion (Fig. 17-1).

   Common dental malocclusions are:

   a. *Esthetic malocclusion*—poor alignment of teeth is esthetically unacceptable in certain cultures.

   b. *Loss of arch continuity*—if a tooth is lost, drifting of adjacent teeth into the space or extrusion of occluding teeth can occur.

   c. *Premature occlusion*—if individual teeth are in heavy occlusion, that is, if they strike one another before the other teeth occlude, excessive forces may be transmitted to periodontal structures. This may cause loosening of the teeth, inflammation of the gingiva, loss of alveolar bone, or in severe instances, temporomandibular joint pain.

   d. *Abnormal cusp relationships*—Irregularities in the intercuspation of teeth occur and chewing efficiency is reduced. Posterior crossbites and end-to-end occlusion are examples.

Notes

18

Directions for Drawing and Carving Teeth

Drawing Teeth

MATERIALS NEEDED

Graph paper ruled 8 squares to the inch
Drawing pencil, 3H or 4H
Eraser
Ruler with millimeter scale
Boley gauge
Teeth

HOW TO DRAW A TOOTH

To make an accurate drawing of any object you have not only to *look* at the object, but also you have to *see* it. A carefully drawn, accurate outline of a tooth is a good indication that you have clearly seen and have understood its morphology. Rarely is there a person, however lacking in artistic skill, who cannot make a reasonably good drawing of an object as simple in outline as a human tooth. Those who are not skilled in *accurate* drawing (extensive art training does not necessarily result in accuracy of outline) may find a solution in using graph paper ruled 8 squares to the inch. The tooth specimen is measured in millimeters with a Boley gauge, and the measurements are transferred to the graph paper allowing 1 square to equal 1 millimeter. Drawings may be made to scale of each type of tooth: maxillary and mandibular incisors, canines, premolars, and molars. Draw maxillary teeth with crowns down; mandibular teeth with crowns up.

Using an undamaged extracted tooth for a specimen, make the following measurements with the Boley gauge (Figs. 18-1 and 18-2).

Crown length	Mesiodistal crown	Faciolingual crown
Root length	Mesiodistal cervix	Faciolingual cervix

On anterior teeth measure the crown length on the facial side from the cervical line to the incisal edge. On posterior teeth measure the crown length on the facial side from the cervical line to the tip of the buccal cusp on premolars, and to the tip of the mesiobuccal cusp on molars. In drawing, make the other cusps their proper length relative to the measured cusp; i.e., either longer or shorter. Using a consistent method of measurement avoids confusion.

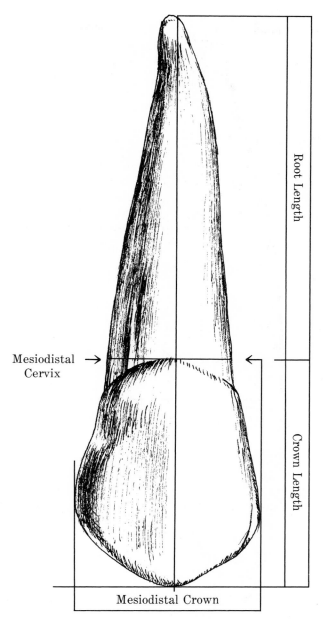

Root Length

Mesiodistal
Cervix →

← Crown Length

Mesiodistal Crown

Fig. 18-1. Drawing of the facial side of a maxillary right canine tooth model to show how measurements of a tooth may be made to assist in drawing and carving.

Plan how you want to place your drawings on the graph paper. One convenient arrangement is: facial aspect, upper left corner; lingual aspect, upper right; mesial aspect, lower left; distal aspect, lower right; incisal aspect, center (Fig. 18-3).

EXAMPLE: How to draw a maxillary central incisor.

Facial Aspect

Use the measurements you have made of the tooth specimen you intend to draw. In the upper left corner of the page, count down from the upper border the number of squares and fraction thereof equal to the root length in millimeters and draw a horizontal line. From

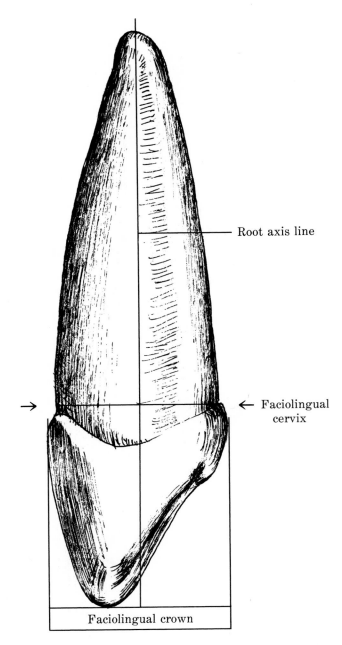

FIG. 18-2. Drawing of the mesial side of the tooth model depicted in figure 18-1 to show how tooth measurements may be made.

this line count down the number of squares equal to the crown length in millimeters and draw a horizontal line. From the left border of the graph paper count to the right the number of squares equal to the mesiodistal crown measurement and draw a vertical line.

Inside this box you will draw the facial aspect of the tooth. Make your lines *very light* at first, so that corrections can be made easily.

Before you start to draw, make a light check mark at the locations of the mesial and distal contact areas of the crown. (A pencil held against the side of the tooth parallel to the root axis will help you determine where to put the check mark.) Also, mark the location of the apex of the root. Estimate the location mesiodistally of the cervix of the tooth. When you fit the crown into the box, if you remember to keep the root vertical, the axis may not be an equal

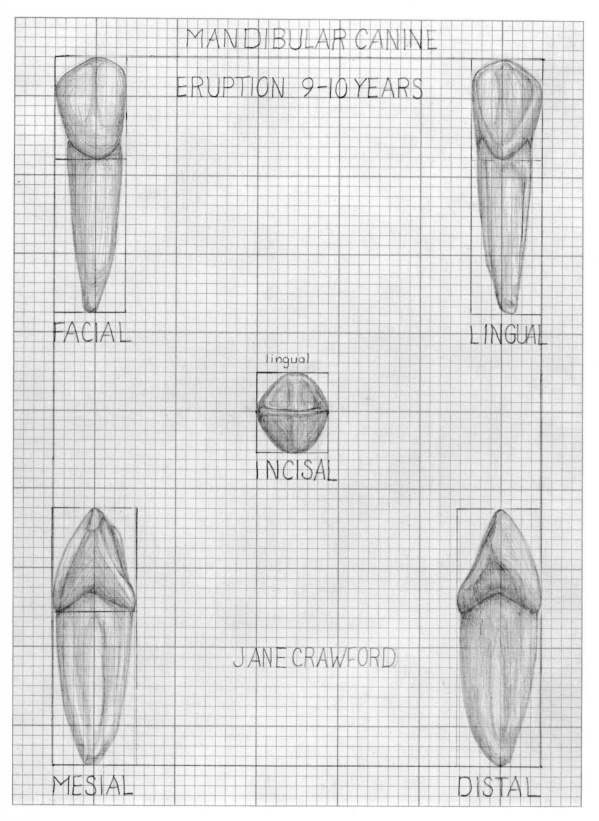

FIG. 18-3. Student drawing of a model of a mandibular right canine. This is one way of drawing teeth to learn their morphology. Shading the drawing is interesting but not necessary.

distance from the mesial and distal sides because the crown of some teeth is tilted distally. Mark off the mesiodistal cervical measurement very lightly.

Now draw in the curvature of the crown at the contact areas (you marked the location), and draw in a sector of the cervical line and a sector of the incisal edge. Draw the root apex, and the cervical part of the root. Correct any errors in location or shape, and then connect the lines you have drawn. You have a tooth.

LINGUAL ASPECT

Use the same set of measurements to make the box in which to draw the lingual aspect of the tooth.

MESIAL AND DISTAL ASPECTS

To draw the boxes, use the root length, the crown length, and the *faciolingual* crown measurement (instead of the mesiodistal measurement). Before you start to draw the tooth, lightly mark the locations of the incisal edge, the labial crest of curvature (i.e., where the curve of the labial surface will touch the line of the box), and the crest of curvature on the cingulum (see Fig. 2-3), and the root tip. Mark the faciolingual width of the cervix. Then draw the tooth.

INCISAL ASPECT

Near the center of the page draw a box with the upper and lower horizontal lines the number of squares equal to the mesiodistal measurement of the crown in millimeters, and the right and left vertical lines the number of squares equal to the faciolingual crown measurement in millimeters. Hold the tooth facial side up and in such a position that you are looking exactly in line with the root axis. Be sure that the tooth is not tilted up or down. On the sides of the box mark the places where you are going to put the mesioincisal and distoincisal angles. The incisal edge of the tooth will be approximately horizontal (it will have a slight curvature) and will lie either in the center of the box faciolingually or slightly facial to the center (whichever is shown on your drawings of the mesial and distal aspects). Watch the position of the cingulum: it is distal to the center of the lower line of the box; and the mesial marginal ridge is longer than the distal marginal ridge.

Do you find any straight lines (ruler-straight, that is) on any tooth other than those lines which have been produced by attrition?

Using this same approach, you will be able to draw other types of teeth. Labeling the grooves, the fossae, and the ridges on the occlusal surfaces of the posterior teeth will help to fix the morphology in your mind (Figs. 6-4, 7-3, 7-7, 8-3, and 9-2).

Carving Teeth

MATERIALS NEEDED

Blocks of carving wax (about 34 × 17 × 17 mm.)
Boley gauge
Millimeter rule
Office knife
Roach carver
Sharp 3H or 4H drawing pencil

HOW TO CARVE A TOOTH

Carving a tooth helps you to see the tooth in 3 dimensions and also to develop considerable manual skill. While eventually you may be able to carve a tooth from a block of wax without preliminary measurement, the beginner will do well to approach the carving in the way he approached the drawing: by drawing a box on the wax block, drawing an outline of the tooth in the box, and then carving around the outline.

Perhaps, after all, it is encouraging rather than ridiculous to approach the task of carving a tooth with the thought in mind that Michelangelo conceived of his task of producing a marble statue as "liberating the figure from the marble that imprisons it." And remember that he, too, sometimes made mistakes and had to discard: a half-finished statue of St. Matthew appears to the casual observer to be all right from the front, but from the side the leg, bent at the knee, is seen to be hopelessly out of position.

The same can happen to your tooth carving. As you cut away, repeatedly examine your carving from all sides; turn it round and round and compare it with your specimen.

EXAMPLE: How to carve a maxillary central incisor (Fig. 18-4).

1. Use the measurements you used for drawing. Again, use the measurement of the buccal cusp on premolars and of the mesiobuccal cusp on molars. This consistency of method prevents confusion. Allowance is made for the greater length of some lingual cusps.

2. Make the sides of the block flat, and all angles right angles.

3. Measure 2 mm. from 1 end of the block and draw a line at this level all around the block. (This end of the block will be the incisal or occlusal end of the tooth and the 2 mm. allowance here is to provide for the extra length of the cusps on molars which are longer than the cusp measured to establish crown length. While it is convenient to allow the 2 mm. on all carvings, it is essential only for molars.)

4. From the 2 mm. line, measure the crown length and draw a line around the block at this level. This second line is the location of the cervical line on the facial side of the tooth.

5. From this cervical line, measure $\frac{1}{2}$ of the length of the root and draw a line around the block. (The end of the block beyond this line will be referred to now as the *base*.)

6. On the base of the block carve, on appropriate sides, *F* (facial), *L* (lingual), *M* (mesial), *D* (distal). Be sure to put the *M* and *D* in the proper relation to *F* and *L* so that you will carve a *right* or a *left* tooth, whichever you intend.

7. Draw a light line lengthwise of the block in the *center* of the mesial surface. Do the same on the distal surface, and BE SURE THAT THESE LINES ARE EXACTLY OPPOSITE.

8. To the figure you have for the faciolingual measurement of the crown, add $\frac{1}{2}$ mm. Divide the resulting figure by 2. Using this latter number, draw a line on either side of the center line that you just drew on the mesial and distal sides of the block (Fig. 18-4a). These lines are to be parallel to the center line and are to extend from the top of the block to the line located at $\frac{1}{2}$ of the root length.

   These lines box in the faciolingual crown dimension plus $\frac{1}{2}$ mm. The $\frac{1}{2}$ mm. is an allowance for safety in carving.

   Do not make trouble for yourself by allowing a large amount (over $\frac{1}{2}$ mm.) of extra measurement.

9. On the surface of the block marked *M*, draw, to fit in the box, an outline of the mesial side of the tooth as you drew it on the graph paper. Be careful to place the incisal edge and the labial and lingual crests of curvature accurately. Your carving will probably be no better than your drawing.

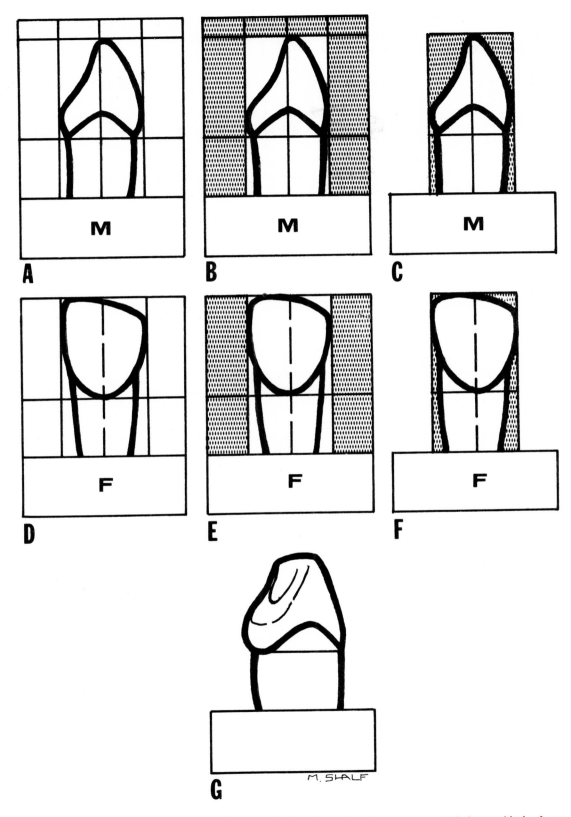

FIG. 18-4. A diagram illustrating the method described in the outline for carving a tooth from a block of wax.

10. Draw a similar outline on the distal side of the block. Be sure that on both sides the drawings are so oriented that the facial surface of the tooth is toward the side of the block you have marked *F*. (It is easy to make a mistake here.)

11. Carve away the excess wax from the facial and lingual sides of the block from the incisal end to the horizontal line that indicates $\frac{1}{2}$ of the root length. *At this time do not carve around the outline of the tooth, but rather carve on the vertical lines that form the box in which the tooth picture is drawn* (Fig. 18-4b).

12. Check these 2 parallel carved surfaces carefully with your Boley gauge. *Be sure they are perfectly flat and smooth. Be sure the thickness of the column of wax between these parallel surfaces exactly equals the given faciolingual crown dimension plus $\frac{1}{2}$ mm.*

13. Now carve around the drawing of the tooth. Follow the drawing carefully, making the tooth shape the same all the way through the block (Fig. 18-4c). *Do not leave a bulge in the center.* Keep the carving surface smooth; if it becomes all chopped up it will be impossible to smooth it without losing both the shape and the size of the carving.

14. Draw center lines, *very lightly,* on the facial and lingual surfaces of the carving. Be sure they are exactly opposite.

15. Add $\frac{1}{2}$ mm. to the mesiodistal crown measurement and draw a line on either side of the center line at a point the distance of $\frac{1}{2}$ of this figure. This makes a box as wide as the greatest mesiodistal crown measurement plus $\frac{1}{2}$ mm.

16. Draw the facial outline of the crown and the root on the facial side of the block (Fig. 18-4d).

17. On the lingual surface of the block draw an outline the same shape as the one on the facial surface except, of course, that it is a mirror image: the distal side of the tooth must be toward the same side of the block in each case.

18. Carve away all the wax outside of the drawing. Leave the base of the block uncarved (Fig. 18-4e).*

19. Shape the tooth by careful carving so that it resembles your tooth specimen from all sides. This operation is one of rounding off the corners, narrowing the lingual surface, shaping the cingulum (it is distal to the center line, and the mesial marginal ridge is longer than the distal), and carving out the lingual fossa. Be sure to look at all aspects of the tooth as you are carving. This includes, of course, the incisal (occlusal) aspect (Figs. 18-4f, g).

20. Carve your initials on the bottom of the base of the block.

*In certain molar teeth the spreading roots may extend beyond the box lines. This must be remembered in carving the roots of these teeth.

Notes

Index